COMPUTATIONAL FUNCTIONAL ANALYSIS

ELLIS HORWOOD SERIES IN MATHEMATICS AND ITS APPLICATIONS

Series Editor: Professor G. M. BELL, Chelsea College, University of London

Statistics and Operational Research

Editor: B. W. CONOLLY, Chelsea College, University of London

Baldock, G. R. & Bridgeman, T. **Mathematical Theory of Wave Motion**
de Barra, G. **Measure Theory and Integration**
Berry, J. S., Burghes, D. N., Huntley, I. D., James, D. J. G. & Moscardini, A. O. **Teaching and Applying Mathematical Modelling**
Burghes, D. N. & Borrie, M. **Modelling with Differential Equations**
Burghes, D. N. & Downs, A. M. **Modern Introduction to Classical Mechanics and Control**
Burghes, D. N. & Graham, A. **Introduction to Control Theory, including Optimal Control**
Burghes, D. N., Huntley, I. & McDonald, J. **Applying Mathematics**
Burghes, D. N. & Wood, A. D. **Mathematical Models in the Social, Management and Life Sciences**
Butkovskiy, A. G. **Green's Functions and Transfer Functions Handbook**
Butkovskiy, A. G. **Structure of Distributed Systems**
Chorlton, F. **Textbook of Dynamics, 2nd Edition**
Chorlton, F. **Vector and Tensor Methods**
Dunning-Davies, J. **Mathematical Methods for Mathematicians, Physical Scientists and Engineers**
Eason, G., Coles, C. W. & Gettinby, G. **Mathematics and Statistics for the Bio-sciences**
Exton, H. **Handbook of Hypergeometric Integrals**
Exton, H. **Multiple Hypergeometric Functions and Applications**
Exton, H. ***q*-Hypergeometric Functions and Applications**
Faux, I. D. & Pratt, M. J. **Computational Geometry for Design and Manufacture**
Firby, P. A. & Gardiner, C. F. **Surface Topology**
Gardiner, C. F. **Modern Algebra**
Gasson, P. C. **Geometry of Spatial Forms**
Goodbody, A. M. **Cartesian Tensors**
Goult, R. J. **Applied Linear Algebra**
Graham, A. **Kronecker Products and Matrix Calculus: with Applications**
Graham, A. **Matrix Theory and Applications for Engineers and Mathematicians**
Griffel, D. H. **Applied Functional Analysis**
Hanyga, A. **Mathematical Theory of Non-linear Elasticity**
Hoskins, R. F. **Generalised Functions**
Hunter, S. C. **Mechanics of Continuous Media, 2nd (Revised) Edition**
Huntley, I. & Johnson, R. M. **Linear and Nonlinear Differential Equations**
Jaswon, M. A. & Rose, M. A. **Crystal Symmetry: The Theory of Colour Crystallography**
Johnson, R. M. **Linear Differential Equations and Difference Equations: A Systems Approach**
Kim, K. H. & Roush, F. W. **Applied Abstract Algebra**
Kosinski, W. **Field Singularities and Wave Analysis in Continuum Mechanics**
Marichev, O. I. **Integral Transforms of Higher Transcendental Functions**
Meek, B. L. & Fairthorne, S. **Using Computers**
Muller-Pfeiffer, E. **Spectral Theory of Ordinary Differential Operators**
Nonweiler, T. R. F. **Computational Mathematics: An Introduction to Numerical Analysis**
Oldknow, A. & Smith, D. **Learning Mathematics with Micros**
Ogden, R. W. **Non-linear Elastic Deformations**
Rankin, R. A. **Modular Forms**
Ratschek, H. & Rokne, Jon **Computer Methods for the Range of Functions**
Scorer, R. S. **Environmental Aerodynamics**
Smith, D. K. **Network Optimisation Practice: A Computational Guide**
Srivastava, H. M. & Karlsson, P. W. **Multiple Gaussian Hypergeometric Series**
Srivastava, H. M. & Manocha, H. L. **A Treatise on Generating Functions**
Sweet, M. V. **Algebra, Geometry and Trigonometry for Science, Engineering and Mathematics Students**
Temperley, H. N. V. & Trevena, D. H. **Liquids and Their Properties**
Temperley, H. N. V. **Graph Theory and Applications**
Thom, R. **Mathematical Models of Morphogenesis**
Thomas, L. C. **Games Theory and Applications**
Townend, M. Stewart **Mathematics in Sport**
Twizell, E. H. **Computational Methods for Partial Differential Equations**
Wheeler, R. F. **Rethinking Mathematical Concepts**
Willmore, T. J. **Total Curvature in Riemannian Geometry**
Willmore, T. J. & Hitchin, N. **Global Riemannian Geometry**

COMPUTATIONAL FUNCTIONAL ANALYSIS

Ramon E. MOORE, Ph.D.
Professor of Mathematics
University of Texas at Arlington, USA

ELLIS HORWOOD LIMITED
Publishers · Chichester
Halsted Press: a division of
JOHN WILEY & SONS
Chichester · New York · Ontario · Brisbane

First published in 1985 by
ELLIS HORWOOD LIMITED
Market Cross House, Cooper Street, Chichester, West Sussex, PO19 1EB, England

The publisher's colophon is reproduced from James Gillison's drawing of the ancient Market Cross, Chichester.

Distributors:
Australia, New Zealand, South-east Asia:
Jacaranda-Wiley Ltd., Jacaranda Press,
JOHN WILEY & SONS INC.,
G.P.O. Box 859, Brisbane, Queensland 40001, Australia
Canada:
JOHN WILEY & SONS CANADA LIMITED
22 Worcester Road, Rexdale, Ontario, Canada.
Europe, Africa:
JOHN WILEY & SONS LIMITED
Baffins Lane, Chichester, West Sussex, England.
North and South America and the rest of the world:
Halsted Press: a division of
JOHN WILEY & SONS
605 Third Avenue, New York, N.Y. 10016, U.S.A.

British Library Cataloguing in Publication Data
Moore, Ramon E.
Computational functional analysis. –
(Ellis Horwood series in mathematics and its applications)
1. Functional analysis
I. Title
515.7 QA320

Library of Congress Card No. 84-25168

ISBN 0-85312-807-3 (Ellis Horwood Limited – Library Edn.)
ISBN 0-85312-814-6 (Ellis Horwood Limited – Student Edn.)
ISBN 0-470-20119-3 (Halsted Press)

Printed in Great Britain by Butler & Tanner, Frome, Somerset.

Contents

for Devin and Claire

Preface

The term 'functional analysis' now refers to a fruitful and diversified branch of mathematics which includes the study of set-theoretic, topological, algebraic, geometric, order, and analytic properties of mappings in finite and infinite dimensional spaces. It is characterized by a generality and elegance which is lacking in classical analysis. Computational mathematics and numerical analysis now rely heavily on results from this theory.

In these lecture notes, the main emphasis is on numerical methods for operator equations – in particular, on the analysis of approximation error in various methods for obtaining approximate solutions to equations and systems of equations. These might be algebraic, linear, non-linear, differential, integral, or other types of equations.

An important part of functional analysis is the extension of techniques for dealing with finite dimensional problems to the infinite dimensional case. This allows us to obtain results which apply at the same time to finite systems of algebraic equations or equally to differential and integral equations.

In mathematics, there is often a trade-off between generality and precision. As a result, in any specific application of functional analysis to a particular numerical problem, there is always the possibility of sharpening results by making use of special properties of the particular problem. In spite of this, the methods of functional analysis are, at the very least, an excellent starting point for any practical problem.

This text is designed for a one-semester introduction at the first year graduate level; however, the material can easily be expanded to fill a two-semester course. It has been taught both ways by the author at the University of Wisconsin-Madison and as a one-semester course at the University of Texas at Arlington. By adding a little additional detail and proceeding at a leisurely pace, Chapters 1–9 and 11–13 can serve as the first semester's material concentrating on *linear* operator equations. The remaining material, concentrating on *nonlinear* operator equations, can serve as the second semester's material, again with a little additional detail and proceeding at a comfortable pace. The material as written can be covered in one semester as a concentrated introduction for students who are willing to work hard to acquire, in a short period, the rudiments of a powerful discipline.

An easy way to expand the material to fit a two-semester course is for the instructor to discuss in detail every one of the more than 100 exercises in the text *after* the students have had a try at them.

It is no more possible to acquire mathematical strength and skills by simply sitting in a lecture room and listening to someone talk about mathematics than it is to acquire physical strength and skills by sitting in a living room and watching football on television. Therefore, it is essential for the education of the students that they try all the exercises, which are designed to help them learn how to discover mathematics for themselves.

The usual practise of numbering equations along with frequent cross-references to equations on distant pages has been dropped in this text as an unnecessary encumbrance.

I am grateful for the helpful suggestions of an anonymous referee, who carefully read the first draft.

CHAPTER 1

Introduction

The outcome of any numerical computation will be a finite set of numbers. The numbers themselves will be finite decimal (or binary) expansions of rational numbers. Nevertheless, such a set of numbers can represent a *function* in many ways: as coefficients of a polynomial; as coefficients of a piecewise polynomial function (for example a spline function); as Fourier coefficients; as left and right hand endpoints of interval coefficients of an interval valued function; as coefficients of each of the components of a vector valued function; as values of a function at a finite set of argument points; etc.

The concepts and techniques of functional analysis we will study will enable us to design and apply methods for the approximate solution of operator equations (differential equations, integral equations, and others). We will be able to compute numerical representations of approximate solutions and numerical estimates of error. Armed with convergence theorems, we will know that, by doing enough computing, we will be able to obtain approximate solutions of any desired accuracy, and know when we have done so.

Since no previous knowledge of functional analysis is assumed here, a number of introductory topics will be discussed at the beginning in order to prepare for discussion of the computational methods.

The literature in functional analysis is now quite extensive, and only a small part of it is presented here – that which seems most immediately relevant to computational problems. This is an introductory study. It is hoped that the reader will be brought along far enough to be able to begin reading the more advanced literature and to apply the techniques to practical problems.

Some knowledge of linear algebra and differential equations will be assumed. Previous study of numerical methods and some experience in computing will help in understanding the applications to be discussed. No background in measure theory is assumed; in fact, we will make scant use of those concepts.

In the first part of the study, we will introduce a number of kinds of topological spaces suitable for investigations of computational methods for solving linear operator equations. These will include Hilbert spaces, Banach spaces, and metric spaces. Linear functionals will play an important role, especially in

Hilbert spaces. In fact, these mappings are the source of the name 'functional analysis'. We will see that the Riesz representation theorem plays an important role in computing when we operate in reproducing kernel Hilbert spaces.

The study of order relations in function spaces leads to important computing methods based on interval valued mappings. We will see how interval analysis fits into the general framework of functional analysis.

In the second part of the study, we will turn our attention to methods for the approximate solution of linear operator equations.

In the third part of the study, we will investigate methods for the approximate solution of nonlinear operator equations.

CHAPTER 2

Linear spaces

We begin with an introduction to some basic concepts and definitions in linear algebra. These are of fundamental importance for linear problems in functional analysis, and are also of importance for many of the methods for nonlinear problems, since these often involve solving a sequence of linear problems related to the nonlinear problem.

The main ideas are these: We can regard real valued functions, defined on a continuum of arguments, as points (or vectors) in the same way as we regard n-tuples of real numbers as points; that is, we can define addition and scalar multiplication. We can take linear combinations. We can form larger or smaller linear spaces containing or contained in them; and we can identify equivalent linear spaces, differing essentially only in notation.

Many numerical methods involve finding approximate solutions to operator equations (for example differential equations or integral equations) in the form of polynomial approximations (or other types of approximations) which can be computed in reasonably simple ways. Often the exact solution cannot be computed at all in finite real time, but can only be approximated as the limit of an infinite sequence of computations.

Thus, for numerical approximation of solutions as well as for theoretical analysis of properties of solutions, linear spaces are indispensable.

The basic properties of relations are introduced in this chapter, since they will be met in many different contexts throughout the subsequent chapters.

An understanding of the material in the exercises will be assumed as the text proceeds.

Definition

A *linear space*, or *vector space*, over the field **R** of real numbers is a set X, of elements called points, or vectors, endowed with the operations of addition and scalar multiplication having the following properties:

(1) $\forall\, x, y \in X$ and $\forall\, a, b \in \mathbf{R}$: [$\forall$ = for all; $\in$ = in the set]

$$x + y \in X,$$

$$\begin{aligned} a\,x &\in X, \\ 1\,x &= x, \\ a(b\,x) &= (a\,b)\,x, \\ (a+b)\,x &= a\,x + b\,x, \\ a\,(x+y) &= a\,x + a\,y; \end{aligned}$$

(2) $(X, +)$ is a commutative group; that is, $\forall\, x, y, z \in X$:

$\exists\ 0 \in X$ such that $0 + x = x$,

$\exists\ (-x) \in X$ such that $x + (-x) = 0$, [$\exists$ = there exists]

$x + y = y + x$,

$x + (y + z) = (x + y) + z$.

Examples

(1) $X = \mathbf{R}$ with addition and scalar multiplication defined in the usual way for real numbers;
(2) $X = E^n$, n-dimensional Euclidean vector space, with componentwise addition and scalar multiplication;
(3) X = polynomials, with real coefficients, of degree not exceeding n, with addition defined by adding coefficients of monomials of the same degree and scalar multiplication defined by multiplication of each coefficient;
(4) all polynomials, with real coefficients, with addition and scalar multiplication as in (3);
(5) continuous real valued functions on $\mathbf{R}$ with pointwise addition and scalar multiplication: $(x + y)\,(t) = x(t) + y(t)$ and $(a\,x)\,(t) = a\,x(t)$;
(6) all real valued functions on $\mathbf{R}$ with addition and scalar multiplication as in (5).

Exercise 1 Can we define addition and scalar multiplication for n-by-n matrices with real coefficients so that they form a linear space? Check all the required properties.

Definition

A *linear manifold*, or *subspace*,† of a linear space X, is a subset Y of X which is algebraically closed‡ under the operations of addition and scalar multiplication for elements of X. Thus, Y is itself a linear space.

Exercise 2 Show that the zero element of a linear space X is also an element of every subspace of X.

---

† $x + y$ and $a\,x$ are in Y for all x and y in Y and all real a.
‡ In a *topological* linear space, a subspace is defined as *closed* linear manifold; see Chapter 5.

Exercise 3 Show that examples (3), (4), and (5) are, respectively, subspaces of examples (4), (5), and (6). Can you find any other subspaces of example (6)?

Definition

Two linear spaces X and Y are *isomorphic* if there is a one-one, *linear* mapping of X onto Y: $m(x+y) = m(x) + m(y)$, $m(a\,x) = a\,m(x)$.

Exercise 4 Show that such a mapping has an inverse which is also linear.

Exercise 5 Let T be an arbitrary set with n distinct elements. Show that the linear space of real valued functions on T with pointwise addition and scalar multiplication is isomorphic to E^n.

NOTE: Unless otherwise stated, all linear spaces considered in this text will be over the real scalar field.

Definition

The *Cartesian product* (or *direct sum*) of two linear spaces X and Y, denoted by $X \times Y$ (or $X \oplus Y$), is the set of ordered pairs (x, y) with $x \in X$ and $y \in Y$, endowed with componentwise addition and scalar multiplication:

$$\begin{aligned}(x, y) + (u, v) &= (x+u, y+v) \\ a(x, y) &= (a\,x, a\,y)\ .\end{aligned}$$

Exercise 6 Show that E^n is isomorphic to $E^{n-1} \times \mathbf{R}$.

Definitions

A *relation*, r, in a set X, is a subset of $X \times X$. If (x, y) belongs to the relation r, we write $x\,r\,y$.

A relation is called *transitive* if

$$\forall x, y, z : x\,r\,y \text{ and } y\,r\,z \text{ implies } x\,r\,z\ .$$

A relation is called *reflexive* if

$$\forall x : x\,r\,x\ .$$

A relation is called *symmetric* if

$$\forall x, y : x\,r\,y \text{ implies } y\,r\,x\ .$$

An *equivalence* relation is a relation that is transitive, reflexive, and symmetric. An equivalence relation in a set X factors X into *equivalence classes*. Denote by C_x the equivalence class to which x belongs. Thus $y \in C_x$ means that $y\,r\,x$.

Exercise 7 Show that two equivalence classes in a set X are either disjoint or they coincide.

Exercise 8 Suppose r is an equivalence relation in a linear space X. Suppose further that $x' \in C_x$ and $y' \in C_y$ imply that $x' + y' \in C_{x+y}$ and $a\,x' \in C_{ax}$ for all real a. Show that the set of equivalence classes is again a linear space with

$$C_x + C_y = C_{x+y}$$

and

$$a\,C_x = C_{ax} \ .$$

Examples

(1) Suppose that Y is a subspace of a linear space X. We can define an equivalence relation in X by

$$\forall x, y \in X \ , \ x\,r\,y \text{ if and only if } x - y \text{ is in } Y.$$

The linear space of equivalence classes defined in this way is called the *factor space, X modulo Y*, written X/Y. The elements of X/Y can be regarded as parallel translations of the subspace Y, since each element of X/Y except for the equivalence class of 0 is disjoint from Y (does not intersect Y). Each element of X/Y is a set in X of the form $x + Y$, that is, the set of all elements of X which are the sum of x and an element of Y.

(2) Let X be the set of all real valued continuous functions on an interval $[a, b]$ in the real line. Let Y be the subspace of functions which vanish at the endpoints a and b. Then X/Y consists of the sets of functions which have given values at a and b.

Exercise 9 Let $X = E^2$ and let Y be a one-dimensional subspace of X. Sketch the elements of X/Y.

CHAPTER 3

Topological spaces

In this chapter, we introduce some basic concepts concerning limits of infinite sequences of points and continuity of functions. The most general setting for these concepts is a topological space. In subsequent chapters, we will consider special kinds of topological spaces which arise most often in computational problems, such as normed linear spaces, inner product spaces, metric spaces, and reproducing kernel Hilbert spaces. Among the metric spaces we will consider are spaces of interval valued functions. All these are topological spaces, so that whatever we can learn about topological spaces will apply to them all.

The concept of compactness plays an important role in the development of methods for finding approximate solutions by numerical methods. We begin, in this chapter, with the most general definition of compactness. While it is equivalent, for most of the spaces used in numerical approximation, to sequential compactness (which will be defined in the next chapter), it is sometimes easier to use the more general definition in theoretical arguments.

Definitions

A *topological space* is a set endowed with a *topology*, which is a family of subsets called *open* sets with the properties:

1) the intersection of any two open sets is an open set,
2) the union of any collection of open sets is an open set, and
3) the empty set and the whole space are open sets.

If Y is a subset of a set X, then the *complement* of Y in X is the set of elements of X which are *not* in Y. We denote the complement of Y in X by $X \setminus Y$.

A subset of a topological space is called *closed* if its complement is open.

An element of a topological space is called a *point*.

A *neighborhood* of a point is a set which contains an open set to which the point belongs.

A point x is a *limit point* of a subset Y of a topological space X if every neighborhood of x contains points of Y other than x.

Let S and T be arbitrary non-empty sets. A *function*, or *mapping*, f, from S into T, denoted by $f: S \to T$, is a set of ordered pairs (s, t) with $s \in S$ and $t \in T$ with the property that if (s, t_1) and (s, t_2) belong to f, then $t_1 = t_2$. If (s, t) belongs to the function f, then t is called the *value*, or *image*, of f at s, denoted by $f(s)$, or f_s.

Let $N = \{1, 2, 3, \ldots.\}$ be the set of counting numbers (non-negative integers). A *sequence* of points in X, $\{x_i\}$, is a mapping of N into X.

Example
If X is a set of functions, then a sequence of 'points' in X is a sequence of functions.

Definition
A sequence of points $\{x_i\}$ in a topological space X *converges* to a point x in X if and only if the sequence eventually lies in any neighborhood of x; that is, given a neighborhood of x, say N_x, there is an integer n such that $x_i \in N_x$ for all $i > n$.

The *closure*, $\bar{Y}$, of a subset Y is the union of Y with all its limit points.

The types of topological spaces we will be mainly concerned with in this text are: metric spaces, normed linear spaces, and inner product spaces. They will be defined and discussed in subsequent chapters. We conclude this chapter with a few more important concepts which apply to all topological spaces.

Definitions
A subset S of a topological space X is *compact* if and only if every open covering of S has a finite subcovering; that is, let F be a family of open sets in X whose union contains S, then S is compact in X if and only if there is a finite collection of elements of F whose union contains S.

A subset S of a topological space X is *relatively compact* if and only if its closure is compact.

A subset S of a topological space X is *sequentially compact* if and only if every sequence of points in S has a convergent subsequence with limit point in S.

Let X and Y be topological spaces. A function $f : X \to Y$ is *continuous* if and only if the inverse image of each open set in Y is an open set in X. (The inverse image of a set S in Y is the set of points in X which map into S.)

Exercise 10 Show that a continuous function maps compact sets onto compact sets; that is, the image of a compact set under a continuous mapping is compact.

Exercise 11 Show that a continuous function maps convergent sequences onto convergent sequences.

Definitions

Let X and Y be topological spaces and let $f: X \to Y$ be a mapping of X into Y. The mapping f is said to be *onto* Y if, for every element y of Y, there is an element x of X such that $f(x) = y$.

The mapping f is said to be a *one-one* mapping if $f(x) = f(z)$ implies $x = z$. Thus a one-one mapping has different images for different points of X. A mapping which is one-one and onto has an *inverse* which is also one-one and onto, denoted by f^{-1}. Thus, $f(f^{-1}(y)) = y$ and $f^{-1}(f(x)) = x$ for all x in X and all y in Y.

We denote the *composition* of two mappings f and g by $f \circ g$; and define it by $(f \circ g)(x) = f(g(x))$.

The *identity* mapping on a space X is the mapping $I_X : X \to X$ such that $I_X(x) = x$, $\forall\, x \in X$. If $f : X \to Y$ is one-one and onto, and if we denote its inverse by $f^{-1}: Y \to X$, then we have $f^{-1} \circ f = I_X$ and $f \circ f^{-1} = I_Y$.

A *homeomorphism* is a mapping $f : X \to Y$ which is continuous, one-one, and onto and has an inverse mapping which is also continuous, one-one, and onto. If X and Y are topological spaces for which there exists a homeomorphism $f : X \to Y$, then X and Y are said to be homeomorphic and called *topologically equivalent* spaces.

Exercise 12 Show that the unit sphere $\{(x, y, z) : x^2 + y^2 + z^2 = 1\}$ in E^3 with the point $(0, 0, 1)$ at the 'North pole' removed is topologically equivalent to E^2.

CHAPTER 4

Metric spaces

A metric space is a particular kind of topological space in which the topology is defined by a distance function. An open set containing a certain point is, for example, the set of all points closer to that point than a given positive number. Actually, an open set in a metric space may be the union of many such sets. In any case, in a metric space, we have the notion of distance between points. In the next two chapters, we will look at two important special cases of metric spaces, namely normed linear spaces and inner product spaces. In these, the distance function is defined by a norm. In the first, any norm; in the second, by a special kind of norm that is derived from an inner product, which is a generalization of inner product (or dot product) of vectors in finite dimensional spaces. The spaces involved in functional analysis are, in general, infinite dimensional.

In a metric space, we can introduce the very important concept of a *Cauchy sequence*. It is any sequence of points that has the property that eventually all its members become close together; that is, the distances between them get small as we go far out in the sequence. When this implies that the limit is in the space, then the space is called complete.

The most important kind of metric space for computational purposes is a complete and separable metric space. Precise definitions are given in this chapter. Important consequences of these definitions are developed in subsequent chapters.

Definition

A *metric space* X is a topological space in which the topology is given by a *metric*, or *distance function*, d, which is a non-negative, real valued mapping of $X \times X$ with the properties:

$\forall x, y, z \in X$:

1) $d(x, y) = 0$ iff $x = y$ [iff = if and only if]
2) $d(x, y) = d(y, x)$, and
3) $d(x, z) \leqslant d(x, y) + d(y, z)$.

A set S in X is *open* iff, $\forall x \in S, \exists r_x > 0$ such that $\{y : d(y, x) < r_x\} \subseteq S$.

The properties required for a metric are those of the ordinary intuitive notion of distance: the distance between two points cannot be negative; the distance between two points is zero if and only if the two points coincide; the distance from a point x to a point y is the same as the distance from y to x; the length of one side of a triangle cannot exceed the sum of lengths of the other two sides. A subset S of X is open if every point in S is in an open sphere contained in S. This means that along with every point it contains, S must also contain the set of all points in X whose distance from x is less than some positive r_x (which may depend on x).

Exercise 13 Show that the definition of a continuous function given in the previous chapter is equivalent, for metric spaces, to the following: if X and Y are metric spaces with metrics d_X and d_Y, respectively, then $f : X \to Y$ is continuous if and only if

$$\forall \epsilon > 0, \ \forall x \in X, \exists \ \delta_{x,\epsilon} > 0 \text{ such that}$$
$$d_X(y, x) < \delta_{x,\epsilon} \text{ implies } d_Y(f(y), f(x)) < \epsilon \ .$$

Exercise 14 Show that a sequence $\{x_i\}$ in a metric space X converges to x in X iff $\lim_{i \to \infty} d(x_i, x) = 0$.

Exercise 15 Show that a subset of a metric space is compact iff it is sequentially compact.

Definition
If X, d_X and Y, d_Y are metric spaces, then a mapping $f : X \to Y$ of X onto Y is called an *isometry* iff $d_X(x, z) = d_Y(f(x), f(z))$ for all x and z in X.

Exercise 16 Show that an isometry is a homeomorphism.

Definition
A sequence of points $\{x_i\}$ in a metric space with metric d is called a *Cauchy sequence* if it has the following property:

$$\forall \epsilon > 0, \exists \text{ an integer } N_\epsilon \text{ such that } d(x_i, x_j) < \epsilon \text{ whenever } i > N_\epsilon \text{ and } j > N_\epsilon.$$

The members of a Cauchy sequence are eventually all close together.

Definition
A metric space is called *complete* iff every Cauchy sequence of points in it converges to a point in the space.

Definition
A subset of a metric space is said to be *dense* if its closure is the whole space. Thus, for example, the real line is the closure of the set of rational numbers, and so the rational numbers are dense in the real line. Every 'real' number is the limit of a sequence of rational numbers.

Definition
A set S is *countable* iff there is a one-one mapping of N onto S.

Definition
If a metric space contains a countable dense subset, then it is called a *separable* metric space. The real line is a separable metric space with $d(x, y) = |x - y|$, since the rational numbers are countable.

Exercise 17 Show that the rational numbers are countable.

Exercise 18 Show that every compact metric space is separable.

Exercise 19 Show that a convergent sequence is necessarily a Cauchy sequence.

A Cauchy sequence need not always be a convergent sequence in a metric space which is not complete. For example, the rational numbers form a metric space with $d(r, s) = |r - s|$ for rational numbers r and s; however, there are Cauchy sequences of rational numbers which do not converge to a rational number.

Exercise 20 Show that the sequence of rational numbers defined recursively by $x_1 = 2$, $x_{i+1} = (x_i + 2/x_i)/2$, $i = 1, 2, \ldots$, does not converge to a rational number even though it is a Cauchy sequence. First show that it is a Cauchy sequence, and then show what it converges to on the real line.

If a metric space X is not complete, we can complete it by adjoining the limits of all its Cauchy sequences. More precisely, we can find a complete metric space X^* such that X is isometric to a dense subset of X^*.

The relation $\{x_i\}\ r\ \{y_i\}$ iff $\lim\limits_{i\to\infty} d(x_i, y_i) = 0$ is an equivalence relation which partitions the set of all Cauchy sequences in X into equivalence classes. If x^* and y^* are two such classes, we define the distance function

$$d_{X^*}(x^*, y^*) = \lim_{i \to \infty} d(x_i, y_i)$$

where $\{x_i\}$ and $\{y_i\}$ are representative sequences from the classes x^* and y^* respectively.

Exercise 21 Show that X^* is a complete metric space and that X is isometric to a dense subset of X^*.

We can, thus, regard a 'real' number as an equivalence class of Cauchy sequences of rational numbers.

CHAPTER 5

Normed linear spaces and Banach spaces

We come now to a special kind of metric space in which the topology defined by a distance function makes the space not only a metric space, but something more. We combine, at this point, the algebraic structure of a linear space with the topological structure of a metric space by means of the concept of a norm. The ordinary idea of distance between points on the real line is an example, namely $|x - y|$.

While all norms in finite dimensional spaces are topologically equivalent (Exercise 24), this is not the case in infinite dimensional spaces of functions. Consequently, some attention must be paid, in infinite dimensional spaces, to the norm in question. Our interpretation of a close approximation will depend on our measure of closeness.

The Weierstrass approximation theorem (referred to in Exercise 25) states that a continuous real valued function on an interval $[a, b]$ can be uniformly approximated to arbitrary accuracy by a polynomial. The possibility of polynomial approximation is extremely important in numerical functional analysis. Thus, the standard norm in the linear space $C[a, b]$ of continuous real valued functions on an interval $[a, b]$ is the 'uniform norm' $\|f\| = \max_{a \leqslant x \leqslant b} |f(x)|$, and distance in $C[a, b]$ is defined by $d(f, y) = \|f - g\|$. Convergence in $C[a, b]$ means uniform convergence of a sequence of continuous functions. This, of course, implies, but is stronger than, pointwise convergence.

A Banach space is a normed linear space which is complete. It contains the limits of all its Cauchy sequences. A normed linear space combines the algebraic structure of a linear space with the topological structure of a metric space.

Definition

A *normed linear space* X is a linear space which is also a metric space with a metric defined by a *norm*, $d(x, y) = \|x - y\|$, where the norm, $\| . \|$, is a non-negative real valued mapping of X with the properties:

1) $\|x\| = 0$ iff $x = 0$,
2) $\|ax\| = |a| \, \|x\|$, for all x in X and all real a,
3) $\|x + y\| \leqslant \|x\| + \|y\|$, for all x and y in X.

Note that the property (2) gives a norm topology more 'structure' than a general metric topology. This property ties together, to some extent, the algebraic and topological structures of a normed linear space. Property (3) is called the triangle inequality.

Examples of normed linear spaces

1) The real line with $\|x\| = |x|$.
2) E^n with $\|x\|_p = (|x_1|^p + \ldots + |x_n|^p)^{1/p}, p \geqslant 1$;

   special cases: for $p = 2$, we have the 'Euclidean' norm,
   for $p = 1$, we have the norm $|x_1| + \ldots + |x_n|$,
   as $p \to \infty$, we obtain the 'max' norm, $\|x\|_\infty = \max_i |x_i|$.

   In all these norms, we have denoted an element of E^n as $x = (x_1, \ldots, x_n)$.
3) $C[a, b]$, the linear space of continuous real valued functions on an interval $[a, b]$ with pointwise addition and scalar multiplication and with $\|x\| = \max_{t \in [a,b]} |x(t)|$.

Exercise 22 Show that the only possible norms in E^1 are of the form $\|x\| = c|x|$ with $c > 0$.

Exercise 23 Show that a norm is necessarily a continuous mapping.

Exercise 24 Show that for any two norms in E^n, say $\|x\|$ and $\|x\|'$, there are positive real numbers k and K such that for all x in E^n, we have

$$k \, \|x\| \leqslant \|x\|' \leqslant K \, \|x\|.$$

Thus, all norms in E^n are topologically equivalent; and convergence in any norm in E^n implies convergence in any other norm for sequences in E^n. Find k and K when $\|x\| = \|x\|_2$ and $\|x\|' = \|x\|_\infty$.

Exercise 25 Show that the space $C[a, b]$ is separable. (Hint: consider polynomials with rational coefficients and use the Weierstrass approximation theorem.)

Definition

A subset S of a normed linear space is *bounded* iff there is a positive real number K such that, $\forall \, x \in S$, $\|x\| \leqslant K$.

Exercise 26 Show that a subset of E^n is compact iff it is closed and bounded.

In an infinite dimensional normed linear space, a subset might be closed and bounded without being compact. For example, let X be the set of bounded sequences of real numbers with $\|x\| = \sup_i |x_i|$, where $x = (x_1, x_2, \ldots)$. The subset $S = \{x : \|x\| = 1\}$ is closed and bounded, but it is not compact, since $\{(1, 0, 0, \ldots), (0, 1, 0, \ldots), (0, 0, 1, 0, \ldots), \ldots\}$ has no convergent subsequence.

Exercise 27 Let $f : X \to Y$ be a mapping of a normed linear space X with norm $\| \cdot \|_X$ onto a normed linear space Y with norm $\| \cdot \|_Y$ such that there exist k and K, positive real numbers, with

$$k \, \|x - y\|_X \leqslant \|f(x) - f(y)\|_Y \leqslant K\|x - y\|_X, \quad \forall x, y \in X.$$

Show that f is a homeomorphism.

Definition
A normed linear space which is complete is called a *Banach space.*

Exercise 28 Show that E^n and $C[a, b]$ are Banach spaces.

Note that the distance function (metric) $d(x, y) = \|x - y\|$ in a normed linear space is *translation invariant*:

$$d(x + z, y + z) = \|x + z - y - z\| = d(x, y) \text{ for all } x \text{ and } y.$$

Such a metric also satisfies

$$d(ax, ay) = |a| \, d(x, y) \text{ for all real } a \text{ and all } x \text{ and } y \text{ in } X.$$

Definition
In a *normed* linear space, a *subspace* is a linear manifold which is *closed* in the norm topology.

Examples
1) Any linear manifold in E^n is a subspace of E^n in any norm on E^n.
2) Let S be any subset of $[a, b]$. The set of continuous functions which vanish on S is a subspace of $C[a, b]$.

Note that the linear manifold of polynomials is *not* a subspace of $C[a, b]$, since there are sequences of polynomials whose limits in the topology of $C[a, b]$ are not polynomials. In fact, *any* element of $C[a, b]$ is the limit of a sequence of polynomials. Furthermore, there are sequences of polynomials which converge pointwise (for each t in $[a, b]$), but not uniformly (that is, not in the topology of $C[a, b]$). Consider, for example, the polynomials $p_n(t) = t^n$ in $C[0, 1]$. As n increases, we have convergence of $p_n(t)$ to 0 for any t except $t = 1$. For $t = 1$, we have convergence to 1. Thus, we have pointwise convergence, but to a discontinuous function.

CHAPTER 6

Inner product spaces and Hilbert spaces

Just as metric spaces are special kinds of topological spaces and normed linear spaces are special kinds of metric spaces, so inner product spaces are special kinds of normed linear spaces. The distance function, or metric, in an inner product space is a generalization of the ordinary distance in a Euclidean space of finite dimension. In an infinite dimensional inner product space of functions, this is usually an integral, which is an infinite dimensional generalization of a finite sum.

Inner product spaces are particularly important for linear operator equations. Many methods for solving such equations involve orthogonal projections into finite dimensional subspaces. Two vectors, whether finite or infinite dimensional, are orthogonal if their inner product is zero. This is an abstraction of the idea of perpendicular vectors in the plane, for instance.

A complete inner product space is called a *Hilbert space*. The theory of measurable functions and the concept of the Lebesgue integral, while of great importance in many mathematical contexts, is of minor importance in numerical computations and will be almost entirely ignored in this book. The integrals required for computing approximations using methods of orthogonal projection are, in practice, almost always integrals of continuous functions, so that the ordinary Riemann integral will suffice. Nevertheless, it is important to understand the concept of a Hilbert space, since it is required for completeness of an inner product space, that is, for the convergence of Cauchy sequences.

In this chapter we will study orthonormal sequences and their use in representations of infinite sequences of approximations to a given element in a Hilbert space.

Definition

An *inner product space* X is a linear space in which there is defined a real valued function, called an *inner product*, on pairs of elements of X. (For linear spaces over the complex field, the inner product is complex valued; however, we will not discuss this more general type of inner product space in these notes.) If x and y are elements of X, we will denote the inner product of x and y by (x, y).

The defining properties of an inner product are as follows:

1) $(x, x) \geqslant 0$, $\forall x \in X$, with $(x, x) = 0$ iff $x = 0$,
2) $(x, y) = (y, x)$, $\forall x, y \in X$,
3) $(ax + by, z) = a(x, z) + b(y, z)$, $\forall x, y, z \in X$ and all real a and b.

An inner product is *bi-linear*. It is linear in the first argument because of property (3) and in the second argument as well because of the symmetry imposed by property (2).

An inner product space is made into a normed linear space by defining the *inner product norm*: $\|x\| = (x, x)^{1/2}$.

Exercise 29 Show that the inner product norm satisfies properties (1) and (2) of the definition of a norm in the previous chapter.

To verify the triangle inequality, one needs the *Cauchy-Schwarz-Buniakowski* (C.S.B.) inequality:

$$|(x, y)| \leqslant (x, x)^{1/2} (y, y)^{1/2}, \forall x, y \in X,$$

with equality iff x and y are linearly dependent.

We can prove the C.S.B. inequality as follows. If $(x, y) = 0$, then the result follows from property (1) of an inner product. Now suppose that (x, y) is not zero. Put $a = (x, y)/|(x, y)|$ and let b be any real number. Then, for all $x, y \in X$, we have

$$0 \leqslant (ax + by, ax + by) = b^2 (y, y) + 2b\,|(x, y)| + (x, x).$$

This quadratic polynomial is non-negative for all real b, so it cannot have two distinct real roots. Thus, the discriminant must be negative or zero. The C.S.B. inequality follows. The case of equality holds iff $ax + by = 0$ for some b. To show the triangle inequality for the inner product norm, we must show that $(x + y, x + y)^{1/2} \leqslant (x, x)^{1/2} + (y, y)^{1/2}$.

Squaring both sides and expanding, we find that we need $(x, x) + (x, y) + (y, x) + (y, y) \leqslant (x, x) + 2(x, x)^{1/2} (y, y)^{1/2} + (y, y)$. Using the symmetry of the inner product, property (2), and the C.S.B. inequality, it can be seen that the above inequality does, indeed, hold for all x and y in X. The triangle inequality follows from the monotonicity of the square root function for positive arguments.

Definition

We say that x is *orthogonal to* y iff $(x, y) = 0$.

A linear manifold in an inner product space which is *closed* (in the topology defined by the inner product norm) is called a *subspace*, by virtue of the definition of a subspace for a normed linear space.

Exercise 30 Show that, in an inner product space, the subset of points which are orthogonal to a given point is a subspace.

Examples of inner product spaces

(1) E^n is an inner product space with the inner product

$$(x, y) = x_1 y_1 + x_2 y_2 + \ldots + x_n y_n.$$

Thus, the Euclidean norm arises from

$$\|x\|_2 = (x_1^2 + x_2^2 + \ldots + x_n^2)^{1/2} = (x, x)^{1/2}.$$

(2) l_2, the linear space of sequences of real numbers which are *square summable*:

$$\{x_i\} \text{ such that } \sum_{i=1}^{\infty} x_i^2 < \infty \text{ with } \{x_i\} + \{y_i\} = \{x_i + y_i\}$$

and with the inner product

$$\left(\{x_i\}, \{y_i\}\right) = \sum_{i=1}^{\infty} x_i y_i$$

and with $a\{x_i\} = \{ax_i\}$ is an inner product space.

(3) We can make the space of continuous real valued functions on $[a, b]$ into an inner product space with the inner product

$$(f, g) = \int_a^b f(t) g(t) \mathrm{d}t \ .$$

Exercise 31 Show that the sequence $\{f_i\}$ with

$$f_i(t) = \begin{cases} (2t)^{i/2} & \text{for } 0 \leqslant t \leqslant 1/2 \\ 1 - (2(1-t))^{i/2} & \text{for } 1/2 \leqslant t \leqslant 1 \end{cases}$$

is a Cauchy sequence for the inner product norm defined in example (3) above. Show that the sequence is pointwise convergent, but not to a continuous limit. Thus, the inner product space of example (3) is not a *complete* inner product space.

Definition

An inner product space which is complete is called a *Hilbert space.*

Exercise 32 Show that the inner product spaces E^n and l_2, defined in examples (1) and (2) above, are Hilbert spaces.

The inner product space in example (3) above is *not* complete (see Exercise 31); however, it can be completed by the process discussed in Chapter 4. The completion is denoted by $\mathcal{L}_2[a, b]$. In order to make $\mathcal{L}_2[a, b]$ an inner product space (with a metric defined by an inner product norm), the ordinary Riemann integral of example (3) must be extended to the more general Lebesgue integral. The elements of $\mathcal{L}_2[a, b]$ can be regarded as equivalence classes of Lebesgue measurable functions. Two functions belong to the same equivalence class if their values differ only on a subset of $[a, b]$ of Lebesgue measure zero. By construction, a representative function in such an equivalence class can be regarded as the limit (in the topology of $\mathcal{L}_2[a, b]$) of a Cauchy sequence of continuous functions. For continuous functions, the Lebesgue integral is the same as the Riemann integral – that is, it gives the same value. While the concepts of Lebesgue measure and Lebesgue integration are important theoretical concepts in functional analysis, they are of little use in computational applications. The reader is referred to other works on functional analysis for further discussion of these concepts.

The Hilbert space l_2 of example (2) above is separable. It contains the countable dense subset consisting of sequences with only finitely many non-zero components and with these components being rational numbers.

Definitions

A *unit vector* in a Hilbert space is an element whose norm is 1.

An *orthonormal* sequence in a Hilbert space is a sequence $\{e_i\}$ of unit vectors which are mutually orthogonal; thus $(e_i, e_j) = 0$ if i differs from j and $(e_i, e_i) = 1$ for all i.

Example

In l_2, we have the orthonormal sequence

$$\begin{aligned} e_1 &= (1, 0, 0, 0, \ldots) \\ e_2 &= (0, 1, 0, 0, \ldots) \\ e_3 &= (0, 0, 1, 0, \ldots) \\ &\ldots\ldots\ldots\ldots \end{aligned}$$

These vectors are linearly independent. We can view l_2 as the infinite dimensional analog of E^n.

Definition

An orthonormal (O.N.) sequence in a Hilbert space is called *complete* iff there is no nonzero vector in the space which is orthogonal to every vector in the sequence. Thus, if $\{e_i\}$ is a complete O. N. sequence and if $(x, e_i) = 0$ for all i, then $x = 0$.

It can be shown that a Hilbert space contains a complete O. N. sequence iff the space is separable. It can also be shown that any separable Hilbert space is isomorphic and isometric to l_2.

The orthonormal sequence $\{e_i\}$ in l_2 given in the example above is complete. Every element of l_2 has a unique representation of the form:

$$x = \sum_{i=1}^{\infty} x_i e_i$$

for $x = (x_1, x_2, \ldots)$.

Recall that the elements of l_2 are 'square-summable' sequences, so that

$$\|x\| = (x, x)^{1/2} = \left(\sum_{i=1}^{\infty} x_i^2\right)^{1/2} \text{ exists for all } x \text{ in } l_2.$$

The set of all polynomials with rational coefficients forms a countable dense subset of $\mathcal{L}_2[a, b]$; thus, this Hilbert space is separable and is, therefore, isomorphic and isometric to l_2. This is a remarkable fact and is the basis for representations of functions by infinite series expansions such as Fourier series and expansions using sequences of orthogonal polynomials. We can obtain an isomorphism between $\mathcal{L}_2$ and l_2 by mapping such a series expansion onto the sequence of coefficients. We will discuss this in more detail later.

Exercise 33 Show that an orthonormal sequence can be formed from the functions

$$(1/(b-a))^{1/2}, \{(2/(b-a))^{1/2} \cos(2k\pi t/(b-a))\},$$
$$\{(2/(b-a))^{1/2} \sin(2k\pi t/(b-a))\}$$

for $k = 1, 2, \ldots$, in $L_2[a, b]$. Verify that these are mutually orthogonal unit vectors.

Exercise 34 Find polynomials of degrees 0, 1, and 2 which are mutually orthogonal unit vectors in $\mathcal{L}_2[-1, 1]$.

Exercise 35 Show that a finite dimensional linear manifold in a Hilbert space is automatically closed (i.e. a subspace). A linear manifold is of dimension n iff there are n linearly independent elements in the manifold such that every point in the manifold is a linear combination of those elements.

Definition

Let H be a Hilbert space and let $x_1, x_2, \ldots, x_n$ be n linearly independent elements of H. (This means that if $x = c_1x_1 + \ldots + c_nx_n = 0$, then $c_1 = c_2 = \ldots = c_n = 0$.) Let M be the subspace of all linear combinations of $x_1, x_2, \ldots, x_n$. We say that M is *spanned* by $x_1, x_2, \ldots, x_n$, or $M = \text{span}\{x_1, x_2, \ldots, x_n\}$.

If x is a point in a Hilbert space and M is a subspace of H, there is a unique point in M which is closest to x in the metric given by the inner product norm in H. Call this closest point x^*. It can be shown that x^* is the *orthogonal projection* of x on M. It is characterized by the orthogonality of $x - x^*$ to every vector in M. In computational applications of functional analysis, the most important projections are onto finite dimensional subspaces. Therefore, we will prove this result in the case of a finite dimensional subspace.

Theorem Let M be a finite dimensional subspace, of a Hilbert space H, spanned by $x_1, x_2, \ldots, x_n$, and let x be an element of H. Then there is a unique element of M which is closest to x (that is, which minimizes $\|x - y\|$ for y in M). The closest point, x^*, can be found by solving a finite system of linear algebraic equations, and the vector $x - x^*$ is orthogonal to every vector in M.

Proof An arbitrary point in M can be written as

$$y = c_1 x_1 + c_2 x_2 + \ldots + c_n x_n$$

for some set of coefficients $c_1, c_2, \ldots, c_n$. If we have

$$(x - x^*, x_i) = 0 \quad \text{for } i = 1, 2, \ldots, n,$$

for some point x^*, then $x - x^*$ is orthogonal to every point in M because of the linearity of the inner product with respect to the second argument. To see whether there is such a point x^* in M, put $x^* = c_1^* x_1 + c_2^* x_2 + \ldots + c_n^* x_n$. Again, using the linearity of the inner product (this time with respect to the first argument), the above system of equations can be put into the matrix form

$$\begin{pmatrix} (x_1, x_1) & (x_2, x_1) & \ldots & (x_n, x_1) \\ (x_1, x_2) & (x_2, x_2) & \ldots & (x_n, x_2) \\ \cdots & \cdots & \cdots & \cdots \\ (x_1, x_n) & (x_2, x_n) & \ldots & (x_n, x_n) \end{pmatrix} \begin{pmatrix} c_1^* \\ c_2^* \\ \cdot \\ c_n^* \end{pmatrix} = \begin{pmatrix} (x, x_1) \\ (x, x_2) \\ \cdot \\ (x, x_n) \end{pmatrix}.$$

The matrix above is non-singular because of the linear independence of $x_1, x_2, \ldots, x_n$. Thus, there is a unique solution for $c_1^*, c_2^*, \ldots, c_n^*$. Now let y be an arbitrary point in M and put $z = y - x^*$. We have $\|x - y\|^2 = \|x - x^* - z\|^2 = \|x - x^*\|^2 - 2(x - x^*, z) + \|z\|^2$. Since z is in M, it is orthogonal to $x - x^*$. Thus the norm $\|x - y\|$ is minimized for $z = 0$ and the theorem is proved.

Exercise 36 Show that the matrix occurring in the above proof is symmetric and positive definite.

Definitions

The set T of all vectors in a Hilbert space H which are orthogonal to a subspace

S is called the *orthogonal complement* of S. The space H is the *direct sum* of S and T, $H = S \oplus T$. This means that every element of H can be written uniquely as the sum of an element of S and an element of T.

Exercise 37 Prove the assertion of the previous sentence.

An orthonormal sequence, if it is complete, is also called an *orthonormal basis*. For an orthonormal basis and an arbitrary vector x, we have the *Parseval identity*

$$\|x\|^2 = (x, x) = \sum_{i=1}^{\infty} (x, e_i)^2$$

where the basis elements are $e_1, e_2, \ldots$. In fact, x has the unique representation

$$x = \sum_{i=1}^{\infty} (x, e_i)e_i$$

in terms of the given basis. If we view the sequence of coefficients in this representation as an element of l_2, then we can see an isometric isomorphism between a Hilbert space with a complete orthonormal sequence and the space l_2.

For any x and y in a separable Hilbert space, we also have the *Parseval relation*, for basis elements $e_1, e_2, \ldots$,

$$(x, y) = \sum_{i=1}^{\infty} (x, e_i)(e_i, y) \quad ,$$

which follows from the orthonormality of the basis elements. The Parseval identity is the special case when $y = x$.

An important special case of the theorem proved in this chapter occurs when $x_1, x_2, \ldots, x_n$ are mutually orthogonal unit vectors. In this case, the matrix, in the linear system to be solved for the coefficients $c_1^*, \ldots, c_n^*$, is the identity matrix, and the system has the immediate solution $c_i^* = (x, x_i)$ for every i, in fact for every n as well. For the orthonormal sequence of Exercise 33, the resulting coefficients are called the *Fourier coefficients* of x, when x is an element of $\mathcal{L}_2[a, b]$. In general, if $x_1, x_2, \ldots$ is an orthonormal basis in a Hilbert space, the inner products (x, x_i) are called the *generalized* Fourier coefficients of x (with respect to the given basis).

The theorem of this chapter has immediate application to 'least squares' approximation, either discrete or continuous, of a given function by linear combinations of a finite set of basis functions.

In approximation theory, it is shown that, for a continuous function on $[-1, 1]$, a good approximation to the best uniform approximation by polynomials of a given degree (or less) can be found by orthogonal projection of x onto the subspace spanned by *Tchebysheff polynomials* of degrees up to the given degree. The resulting approximation is, at worst, one decimal place less accurate than the best uniform approximation for degrees up to 400.

For details of these and other applications to approximation theory, the reader is referred to a text such as: E. W. Cheney, *Introduction to approximation theory*, McGraw-Hill, 1966, second edition: Chelsea, 1982.

Suppose we have a sequence of vectors $\{y_i\}$ in a Hilbert space H such that the first n members are linearly independent for every n. (For example, the monomials $y_i(t) = t^i$ in $\mathcal{L}_2[a, b]$.) We can construct an orthonormal sequence in H by the *Gram–Schmidt process*: Denote the orthogonal projection of a vector y onto the span of a nonzero vector x by $p(y, x) = [(y, x)/(x, x)]x$. Note that x is orthogonal to $y - p(y, x)$, since

$$(y - p(y, x), x) = (y, x) - [(y, x)/(x, x)]\,(x, x) = 0.$$

Furthermore, $y - p(y, x)$ is again a non-zero vector if x and y are linearly independent. By induction, the vectors

$$\begin{aligned}
z_1 &= y_1 \\
z_2 &= y_2 - p(y_2, z_1) \\
z_3 &= y_3 - p(y_3, z_1) - p(y_3, z_2) \\
&\cdots\cdots\cdots\cdots\cdots \\
z_n &= y_n - p(y_n, z_1) - p(y_n, z_2) - \cdots\cdots - p(y_n, z_{n-1}) \\
&\cdots\cdots\cdots\cdots\cdots\cdots\cdots\cdots
\end{aligned}$$

are mutually orthogonal and nonzero. The sequence of vectors $x_1, x_2, \ldots$ defined by $x_i = z_i/(z_i, z_i)^{1/2}$ is orthonormal.

Exercise 38 Carry out the Gram–Schmidt process for $y_1(t) = 1$, $y_2(t) = t$, $y_3(t) = t^2$ in $\mathcal{L}_2\,[0, 1]$; that is, find z_1, z_2, z_3 and x_1, x_2, x_3 as defined above, using the appropriate inner product.

In summary, if $\{x_i\}$ is a complete orthonormal sequence in a Hilbert space H, then any x in H has a unique representation of the form

$$x = \sum_{i=1}^{\infty} c_i x_i.$$

The (generalized) Fourier coefficients are given by $c_i = (x, x_i)$. The sequence of partial sums converges in the inner product topology to x. (This does *not necessarily* imply pointwise convergence in case the elements of H are functions; but there *are* Hilbert spaces for which the implication *does* hold, as we shall see later.) For every x in H, we have

$$\|x - \sum_{i=1}^{N} (x, x_i)x_i\| \to 0 \text{ as } N \to \infty .$$

Furthermore, for all N and all real $a_1, a_2, \ldots, a_N$, we have

$$\|x - \sum_{i=1}^{N} (x, x_i)x_i\| \leqslant \|x - \sum_{i=1}^{N} a_i x_i\| ;$$

thus, the partial sums with coefficients $(x_1 x_i)$ *are* the 'least squares' approximations to x for every N.

Exercise 39 Find the least squares approximation to e^t on [0, 1] among polynomials in t of degree two or less; that is, find numerical values of a_1, a_2, and a_3 such that

$$\int_0^1 (e^t - (a_1 + a_2 t + a_3 t^2))^2 \, dt$$

is minimum. Graph the resulting polynomial and e^t together, for comparison.

CHAPTER 7

Linear functionals

Linear functionals are real valued functions defined on normed linear spaces (we do not consider the more general complex normed linear spaces in this introductory text). These include evaluation of functions at a given point, finite sums of values, integrals, and inner products. All these are of fundamental importance in computational methods for solving operator equations.

In fact, the set of bounded linear functionals forms a linear space itself, called the conjugate space (or the dual space) of a given normed linear space.

The Riesz representation theorem for bounded linear functionals in Hilbert spaces is the basis for the projection methods discussed in Chapter 9.

Many of the types of convergence discussed in Chapter 8 depend on measures of approximation computed in the dual space.

The term 'functional analysis' itself owes its origins to the study of functionals.

Definitions

A real valued mapping defined on a normed linear space is called a *functional*. If the mapping is linear, it is called a *linear functional*. Thus, $f : X \to \mathbf{R}$ is a linear functional on the normed linear space X iff $f(ax + by) = af(x) + bf(y)$ for all x and y in X and all real a and b.

A linear functional $f : X \to \mathbf{R}$ is *bounded* iff there is a real number B such that $|f(x)| \leqslant B \|x\|$ for all x in X.

Exercise 40 Show that a linear functional is bounded if and only if it is continuous. (Hint: First show that a linear functional is continuous at all x in X iff it is continuous at $x = 0$.)

Examples of linear functionals

(1) *Evaluation functionals*: If the elements of X are themselves functions $x : D \to \mathbf{R}$ on some set D, then we can choose an element of D, say t, and evaluate x at t to obtain $x(t)$, where x is any element of X. We denote this

functional by $\delta_t: X \to \mathbf{R}$. Thus, for any x in X, we have $\delta_t(x) = x(t)$. If X is a linear space of functions with pointwise addition and scalar multiplication then δ_t is a linear functional, since $\delta_t(ax + by) = (ax + by)(t) = ax(t) + by(t) = a\,\delta_t(x) + b\,\delta_t(y)$.

(2) *Finite sums*: Let $w_1, w_2, \ldots, w_n$ be any real numbers and let X be a linear space of functions; then the finite sum

$$s(x) = \sum_{k=1}^{n} w_k x(t_k)$$

is a linear functional on X for any choice of the arguments $t_1, t_2, \ldots, t_n$. We can view this linear functional as a linear combination of evaluation functionals.

(3) If X is a linear space of real valued integrable functions on $[a, b]$, then the *definite integral*

$$I(x) = \int_a^b x(t)\,\mathrm{d}t$$

is a linear functional on X. We have $I(ax + by) = aI(x) + bI(y)$.

(4) *Inner products*: If X is an inner product space and h is any element of X, then $p_h(x) = (x, h)$ is a linear functional on X.

For $X = C[a, b]$, the first three examples above are bounded linear functionals. In fact,

$$|\delta_t(x)| \leqslant \|x\| \ ,$$

$$|s(x)| \leqslant \left(\sum_{k=1}^{n} |w_k| \right) \|x\| \ , \quad \text{and}$$

$$|I(x)| \leqslant |b - a| \, \|x\| \ .$$

Exercise 41 Show that p_h, in example (4) above, is a bounded linear functional.

We have the following important result for bounded linear functionals on Hilbert spaces:

The Riesz Representation theorem If f is a bounded linear functional on a Hilbert space H, then there is a unique element h_f in H, called the *representer* of f, such that $f(x) = (x, h_f)$ for all x in X.

Proof The set $N(f)$ of vectors x in H for which $f(x) = 0$ is called the *null space* of f; it is a subspace of H. If $N(f) = H$, we can take $h_f = 0$. If $N(f)$ is a proper subset of H, there exists a nonzero vector in the orthogonal complement of $N(f)$, in fact a unit vector. Let h be such an element of H. We will now show that $h_f = f(h)h$ satisfies the conclusion of the theorem. To see this, note that $f(x)h - f(h)x$ is in $N(f)$ for all x in H, because

$$f(f(x)h - f(h)x) = f(x)f(h) - f(h)f(x) = 0.$$

Thus, $$(h, f(x)h - f(h)x) = 0,$$

so $$f(x) = (h, f(h)x).$$

It follows that $f(x) = (x, f(h)h)$ for all x in H.

For uniqueness, if $(x, h) = (x, q)$ for all x in H, then in particular, for $x = h - q$, we have $(h - q, h - q) = 0$; so $h = q$. This completes the proof.

Definition
If f is a bounded linear functional on a normed linear space, then

$$\sup_{x \neq 0} \left\{ \frac{|f(x)|}{\|x\|} \right\} = \sup_{\|x\|=1} |f(x)|$$

is called the *norm* of f and is denoted by $\|f\|$.

Exercise 42 Show that, in a Hilbert space, the norm of a bounded linear functional is the norm of its representer.

The following important theorem can be proved for normed linear spaces. We will give the proof in the special case of a Hilbert space only.

The Hahn–Banach theorem A bounded linear functional f_0 defined on a subspace S of a normed linear space X can be extended to X with preservation of norm; that is, if

$$f_0 : S \to \mathbf{R}$$

has norm $\|f_0\|$ in S, then there is a bounded linear functional $f : X \to \mathbf{R}$ such that $f(x) = f_0(x)$ for x in S and $\|f\|$ in X is the same as $\|f_0\|$ in S.

Proof For a Hilbert space H, we have, because of the Riesz representation theorem, $f_0(x) = (x, h)$ for some h in S. We can define $f(x) = (x, h)$ for all x in H; and the theorem is proved (when X is a Hilbert space).

Because of the Hahn–Banach theorem, we can regard any bounded linear functional as defined on the whole of a normed linear space. There are linear functionals which are *unbounded* on a normed linear space. Usually these are defined only on a dense subset of the space. For example,

$$f(x) = \mathrm{d}x/\mathrm{d}t\,|_{t=t_0}, \quad t_0 \in [a, b],$$

is an unbounded linear functional defined on the subset of differentiable functions in $C[a, b]$. By the Weierstrass theorem on polynomial approximation in $C[a, b]$, the polynomials are dense in $C[a, b]$; and, therefore, so are the differentiable functions.

Definition
Let X be a normed linear space. The *conjugate space* (or *dual space*) of X, denoted by X^*, is the normed linear space of all bounded linear functionals on X. If f_1 and f_2 are bounded linear functionals on X (that is, elements of X^*) we define $f_1 + f_2$ and af_1 by

$$(f_1 + f_2)(x) = f_1(x) + f_2(x), \text{ and } (af_1)(x) = af_1(x) \text{ for all } x \text{ in } X.$$

Exercise 43 Show that the definition of a norm for a bounded linear functional given previously satisfies the properties for a norm in X^*.

It can be shown that the conjugate space X^* is always complete, whether X is or not.

Exercise 44 Show that, for a Hilbert space H, the conjugate space H^* is isometrically isomorphic to H.

Exercise 45 Show that every linear functional on E^n is bounded.

CHAPTER 8

Types of convergence in function spaces

For any particular kind of norm, we may consider convergence in that norm of a sequence of elements (usually functions obtained as approximate solutions to some operator equation) to a function of interest (say a solution of an operator equation). Each type of norm constitutes a type of measure of the size of an element. Thus, if, with a certain norm, the norm of the difference between the kth element of a sequence of approximations and the limit of the sequence goes to zero, we have convergence of the sequence to the limiting element. In this chapter, we consider a number of different types of convergence, depending on the measure of approximation or the type of norm used to measure it. The notion of 'weak' convergence involves the dual space. We can identify at least six different types of convergence. There are numerous relations among these types of convergence, and we discuss a few of them in the exercises, which are of particular interest in numerical applications of functional analysis.

Let D be an arbitrary non-empty set; let X be a normed linear space, with norm $\|.\|_X$; and let F be a normed linear space of functions mapping D into X, with norm $\|.\|_F$. Let $f \in F$ and let $\{f_k\}$ be a sequence of functions in F.

We define (along with the usual practise in functional analysis) the following *types of convergence* of the sequence $\{f_k\}$ to f, as $k \to \infty$:

1. *Strong convergence* (convergence in the norm $\|.\|_F$):

$$\|f_k - f\|_F \to 0;$$

2. *Weak convergence* (convergence in the dual space):

$$|g(f_k) - g(f)| \to 0 \text{ for every } g \text{ in } F^* ;$$

3. *Pointwise convergence*:

$$\|f_k(t) - f(t)\|_X \to 0 \quad \text{for all } t \text{ in } D;$$

4. *Uniform* (pointwise) *convergence*:

$$\|f_k(t) - f(t)\|_X \text{ converges } \textit{uniformly} \text{ to 0 for all } t \text{ in } D;$$

In addition, the following types of convergence of sequences in F^* are defined:

5. *–convergence ("*star*" *convergence*) of $\{g_k\}$ to g in F^*:

$\|g_k - g\|_{F^*} \to 0$; and, finally,

6. *weak* **–convergence* (weak convergence in F^*):

$|g_k(f) - g(f)| \to 0$ for every f in F, with

$\|g_k\| \leqslant M$ for all k and some M.

There are various relations among these types of convergence. Some of them are as follows.

Exercise 46 Show that $\{f_k\}$ converges weakly to f, if $\{f_k\}$ is uniformly bounded ($\|f_k\|_F \leqslant M$ for all k) and if $\{f_k\}$ converges weakly to f in some dense subset of F^*.

Exercise 47 Show that the set of all finite sums

$$s(x) = \sum_{k=1}^{n} w_k x(t_k)$$

is dense in $C[0, 1]^*$ and that weak convergence of a sequence in $C[0, 1]$ means that the sequence is uniformly bounded and pointwise convergent.

Exercise 48 Show that strong convergence in $C[0, 1]$ implies uniform convergence, and that weak convergence does not imply strong convergence in $C[0, 1]$.

Exercise 49 Show that a sequence $\{x_k\}$ of functions in $\mathcal{L}_2\,[0, 1]$ which converges weakly to x in $\mathcal{L}_2\,[0, 1]$ and is such that $\|x_k\| \to \|x\|$, converges strongly in $\mathcal{L}_2\,[0, 1]$, that is,

$$\int_0^1 |x_k(t) - x(t)|^2 \, \mathrm{d}t \to 0.$$

Exercise 50 Show that convergence always implies weak convergence and *– convergence always implies weak *–convergence.

Other relations among the various types of convergence depend on details of the norms in X and F.

Exercise 51 Show that, in a Hilbert space, convergence always implies weak convergence, but not necessarily conversely.

Exercise 52 Show that if a sequence converges weakly in a Hilbert space, then the sequence is uniformly bounded.

In the next chapter, we study an interesting and useful type of Hilbert space in which convergence implies pointwise convergence.

CHAPTER 9

Reproducing kernel Hilbert spaces

In this chapter we meet, for the first time in this text, a numerical application of the theory we have been developing. Here we introduce an even more special type of topological space than any we have considered before. It is not only a metric space, but also a normed linear space, an inner product space, even a Hilbert space. Even more than that, it is a special type of Hilbert space in which convergence in the Hilbert space norm implies pointwise convergence. Examples of such spaces are given.

We consider the problem of approximating an integral by finite sums with the error measured in the norm in such a space. We can make use of a number of theoretical tools developed in previous chapters, in particular the Riesz representation theorem and orthogonal projection.

In the exercises in this chapter, the reader has his first opportunity in this text to taste the flavor of *numerical functional analysis*.

The ambitious student may try to do Exercise 57 for arbitrary n.

Definition
A Hilbert space H of real valued functions on a set D is called a *reproducing kernel Hilbert* (R.K.H.) *space* iff all the evaluation functionals on H are continuous (bounded).

Theorem In a Hilbert space of real valued functions on a set D, convergence implies pointwise convergence iff all the evaluation functionals are continuous (that is, iff the space is an R.K.H.S.).

Proof If δ_t is bounded, then (by the Riesz representation theorem) there is an element R_t in H such that

$$\delta_t(x) = x(t) = (x, R_t) \text{ for all } x \text{ in } H \text{ and all } t \text{ in } D.$$

Now let $\{x_k\}$ be a sequence in H and let x be an element of H. Then, for all t in D, we have

$$|x_k(t) - x(t)| = |(x_k(t) - x(t), R_t)| \leqslant \|R_t\| \, \|x_k - x\| \ .$$

Thus, convergence in H implies pointwise convergence. Conversely, if $\|x_k - x\| \to 0$ implies $|x_k(t) - x(t)| \to 0$ for all t in D, then

$$|\delta_t(x - x_k)| = |x(t) - x_k(t)| \to 0$$

whenever $\|x - x_k\| \to 0$; so δ_t is continuous. This completes the proof.

In a reproducing kernel Hilbert space, we denote the representer of evaluation at t by R_t as in the proof above. As we will show in this chapter, we can find such spaces with *known* functions R_t. In this case, if f is *any* bounded linear functional on H (an R.K.H.S.) then

$$f(x) = (x, h_f) = (h_f, x) ;$$

so

$$f(R_t) = (h_f, R_t) = \delta_t(h_f) = h_f(t).$$

In other words, given the *reproducing kernel* R_t, we can construct the representer of any bounded linear functional as a function of t, by evaluating the linear functional on the kernel R_t.

This elegant theory has important practical applications. It has advantages over the 'Dirac delta function' commonly used in mathematical physics, and also over the theory of distributions (test functions).

Next, we will discuss in detail one example of a reproducing kernel Hilbert space with applications to *interpolation* and *approximation* of integrals. At the end of the chapter we will give other examples of such spaces. Additional applications will occur in later chapters.

We consider now the R.K.H.S. of real valued functions on $[0, 1]$ which are absolutely continuous (indefinite integrals of their derivatives) and whose derivatives are in $\mathcal{L}_2[0, 1]$, with the inner product

$$(f, g) = f(0)g(0) + \int_0^1 f'(s)g'(s)\mathrm{d}s .$$

We denote this Hilbert space by $H^{(1)}$. It includes, of course, all real valued functions on $[0, 1]$ which are continuously differentiable, as well as those which are only piecewise differentiable. The *reproducting kernel* is given by

$$R_t(s) = 1 + \min(s, t), \quad (\text{note that } R_t(0) = 1 \text{ for } t \in [0, 1]),$$

since we have

$$\begin{aligned}(f, R_t) &= f(0) \cdot 1 + \int_0^1 f'(s)R_t'(s)\mathrm{d}s \\ &= f(0) + \int_0^t f'(s)\mathrm{d}s \\ &= f(t) , \qquad \text{for all } t \text{ in } [0, 1] .\end{aligned}$$

It is interesting that we can find, by *orthogonal projection*, a piecewise linear function $\bar{f}$ in $H^{(1)}$ *which interpolates* the values of a given function in $H^{(1)}$ at given arguments $0 \leqslant t_1 < t_2 \ldots < t_n \leqslant 1$. to do this, we put

$$\bar{f} = \sum_{j=1}^{n} c_j R_{t_j} \; .$$

Since $R_{t_j}(s)$ is piecewise linear in s for each t_j, it follows that $\bar{f}(s)$ is also piecewise linear in s. The orthogonal projection $\bar{f}^*$ of an element f of $H^{(1)}$ on the linear manifold spanned by the representers of evaluation at the given arguments $t_1, \ldots, t_n$ is characterized by the system of equations

$$(\bar{f}^* - f, R_{t_i}) = 0, \; i = 1, 2, \ldots, n.$$

Substituting the above expression for $\bar{f}$ and expanding the inner product, we obtain the following system of equations for the coefficients $c_1^*, c_2^*, \ldots, c_n^*$ of $\bar{f}^*$:

$$\sum_{j=1}^{n} (R_{t_j}, R_{t_i}) c_j^* = (f, R_{t_i}) = f(t_i) \; , \; i = 1, 2, \ldots, n.$$

Since we also have

$$\bar{f}^*(t_i) = \sum_{j=1}^{n} c_j^* (R_{t_j}, R_{t_i}) = f(t_i) \; ,$$

it follows that $\bar{f}$ interpolates f at the given arguments. Note that $\bar{f}^*$ minimizes

$$\|f - \bar{f}\|_{H^{(1)}} = \left\{ (f(0) - \bar{f}(0))^2 + \int_0^1 (f'(s) - \bar{f}'(s))^2 \, ds \right\}^{1/2} .$$

Exercise 53 Find $\bar{f}^*$ explicitly for $n = 2$, $t_1 = \frac{1}{3}$, $t_2 = \frac{2}{3}$, $f(t_1) = 2$, $f(t_2) = 1$. Graph the result.

Suppose next that we wish to approximate the definite integral

$$L(f) = \int_0^1 f(s) ds, \;\; \text{for } f \text{ in } H^{(1)} \, ,$$

by a finite sum of the form

$$S_p(f) = \sum_{k=1}^{n} w_k f(t_k)$$

where $p = (w_1, w_2, \ldots, w_n, t_1, t_2, \ldots, t_n)$ is a $2n$-dimensional vector of real parameters with $0 \leqslant t_1 < t_2 < \ldots < t_n \leqslant 1$. Both L and S_p are bounded linear functionals on $H^{(1)}$.

Exercise 54 Prove the assertion of the previous sentence. (Hint: show that L is continuous at $f = 0$.)

We define the *error functional* E_p by $E_p(f) = L(f) - S_p(f)$.

We have $\quad E_p(f) = (f, h_{E_p}) = (f, h_L) - (f, h_{S_p})$,

and so $\quad h_{E_p} = h_L - h_{S_p}$.

We have $\quad |E_p(f)| \leqslant \|E_p\| \, \|f\|$ for all f in $H^{(1)}$;

thus, to minimize $\|E_p\|$ we minimize $\|h_{E_p}\|$ since these are the same.
We find that

$$h_L(t) = L(R_t) = \int_0^1 (1 + \min(s, t))\, ds = 1 + t(1 - t/2)$$

is the representer of the definite integral and

$$h_{S_p}(t) = S_p(R_t) = \sum_{R=1}^{n} w_k R_t(t_k) = \sum_{R=1}^{n} w_k R_{t_k}(t)$$

is the representer of the finite sum.

There are two problems to consider:

(1) For a *given* choice of $t_1, t_2, \ldots, t_n$ to minimize $\|E_p\|$ among all choices of $w_1, \ldots, w_n$. This is a linear approximation problem.
(2) For fixed n, minimize $\|E_p\|$ among all choices of $t_1, \ldots, t_n$ (using the best $w_1, \ldots, w_n$ for each n-tuple $(t_1, \ldots, t_n)$). This is a nonlinear approximation problem. We do *not* have $\min(t, t_k) + \min(t, t_j) = \min(t, t_k + t_j)$.

We can solve the linear problem as follows. We minimize

$$\|E_p\| = \|h_{E_p}\| = \|h_L - h_{S_p}\|$$

by finding the orthogonal projection of h_L on the finite dimensional linear subspace spanned by the representers of evaluation at $t_1, \ldots, t_n$. To do this, we put

$$(h_L - \sum_{k=1}^{n} w_k R_{t_k}, R_{t_j}) = 0\,, \; j = 1, 2, \ldots, n.$$

Thus, the best $w_1, \ldots, w_n$ are found by solving the linear system

$$\sum_{k=1}^{n} R_{t_k}(t_j)\, w_k = h_L(t_j)\,, \; j = 1, 2, \ldots, n\,.$$

More explicitly, the system has the form

$$\sum_{k=1}^{n} (1 + \min(t_k, t_j))\, w_k = 1 + t_j(1 - t_j/2), \; j = 1, 2, \ldots, n\,.$$

Thus, the solution of problem (1) is given by those values of $w_1, w_2, \ldots, w_n$ for which $h_{S_p}(t)$ interpolates $h_L(t)$ at the given values of t. For any particular set of distinct values $0 \leqslant t_1 < t_2 < \ldots < t_n \leqslant 1$, we can find the values of $w_1, \ldots, w_n$, using a numerical method for solving the linear algebraic system (see Exercise 53, for example).

The solution of problem (2) is more difficult. To begin with, it will be helpful to find a general solution of problem (1) giving the w_i's as functions of the t's. To accomplish this, it is convenient to first orthonormalize the basis functions R_{t_k}. We can do this using the Gram–Schmidt process. For $n = 2$, we find that

$$w_1 = t_2/2 + t_1/[2(1 + t_1)] \quad \text{and}$$
$$w_2 = 1 - (t_1 + t_2)/2$$

by direct computation without the Gram–Schmidt process. For n greater than 2, we proceed as follows. We put

$$e_1(t) = R_{t_1}(t)/\|R_{t_1}\| = (1 + \min(s, t_1))/(1 + t_1)^{1/2}.$$

Using the Gram–Schmidt process, we find that, for $k = 2, \ldots, n$,

$$e_k(t) = (t_k - t_{k-1})^{-1/2} \cdot \begin{cases} 0, \text{ for } t \leqslant t_{k-1}, \\ t - t_{k-1}, \text{ for } t_{k-1} \leqslant t \leqslant t_k, \\ t_k - t_{k-1}, \text{ for } t_k \leqslant t. \end{cases}$$

Conversely, we can express the representers of evaluation in terms of these e's as follows

$$R_{t_1}(t) = (1 + t_1)^{1/2} e_1(t)$$
$$R_{t_2}(t) = (t_2 - t_1)^{1/2} e_2(t) + (1 + t_1)^{1/2} e_1(t)$$
$$\ldots\ldots\ldots\ldots\ldots\ldots\ldots\ldots\ldots\ldots\ldots$$
$$R_{t_n}(t) = (t_n - t_{n-1})^{1/2} e_n(t) + \ldots + (t_2 - t_1)^{1/2} e_2(t) + (1 + t_1)^{1/2} e_1(t).$$

With the help of thos orthonormal basis for the linear manifold of approximations, we obtain an upper triangular linear algebraic system to solve for the w's of the form

$$(1 + t_1)^{1/2} (w_1 + w_2 + \ldots + w_n) = (h_L, e_1)$$
$$(t_2 - t_1)^{1/2} (\quad w_2 + \ldots + w_n) = (h_L, e_2)$$
$$\ldots\ldots\ldots\ldots\ldots\ldots\ldots\ldots\ldots\ldots\ldots$$
$$(t_n - t_{n-1})^{1/2} (\quad w_n) = (h_L, e_n).$$

We find that

$$(h_L, e_1) = [1 + t_1(1 - t_1/2)]/(1 + t_1)^{1/2}$$
$$(h_L, e_2) = (t_2 - t_1)^{1/2}\ [1 - (t_2 + t_1)]$$
$$\ldots\ldots\ldots\ldots\ldots\ldots$$
$$(h_L, e_n) = (t_n - t_{n-1})^{1/2}\ [1 - (t_n + t_{n-1})]\ .$$

Thus, we find that, for $n \geqslant 3$,

$$\begin{aligned} w_n &= 1 - (t_n + t_{n-1})/2 \\ w_{n-1} &= (t_n - t_{n-2})/2 \\ w_{n-2} &= (t_{n-1} - t_{n-3})/2 \\ &\ldots\ldots\ldots \\ w_2 &= (t_3 - t_1)/2 \\ w_1 &= [t_1/(1 + t_1) + t_2]/2\ . \end{aligned}$$

Recall that, for the case $n = 2$, the first and last of the above expressions give the correct result.

With these formulas, we can express $\|E_p\| = \|h_{E_p}\| = \|h_L - h_{S_p}\|$ as a function of $t_1, \ldots, t_n$.

We find, that after some elementary algebra, that

$$\|h_{E_p}\|^2 = (1/3)\,(1 - t_n)^3 + (2/3)\,\{((t_n - t_{n-1})/2)^3 + \ldots + ((t_2 - t_1)/2)^3\}$$
$$+ (1/3)t_1^3\,(1 - 3t_1/(4 + 4t_1))\ ,\ \text{for } n \geqslant 3.$$

We will minimize $\|h_{E_p}\|$ among all admissible choices of

$$0 \leqslant t_1 < t_2 \ldots < t_n \leqslant 1$$

if we minimize $\|h_{E_p}\|^2$.

We will not solve this minimization problem here. However, it is clear that for any admissible choices of the t_k's we can evaluate the upper bound on the minimum possible value of $\|h_{E_p}\|$ using the formula above. For example, with

$$t_1 = 0,\ t_n = 1,\ \text{and}\ t_k = (k - 1)/(n - 1),\ k = 1, 2, \ldots, n,$$

we find that $\|h_{E_p}\| = (1/2\sqrt{3})/(n - 1)$.

Exercise 55 Conclude that we have proved from the above analysis that, for all f in $H^{(1)}$, $S_p(f)$ converges to $L(f)$ when the w's and t's are chosen as described (either equally spaced t's or optimally spaced t's).

Exercise 56 Show that, for equally spaced t's, with $t_1 = 0$ and $t_n = 1$, piecewise linear interpolation of a given function f in $H^{(1)}$ by orthogonal projection on the subspace spanned by the representers of evaluation at the t's converges, as $n \to \infty$, in the $H^{(1)}$ norm and, hence, pointwise to the given function.

Exercise 57 For $n = 2$, solve both minimization problems (1) and (2) explicitly.

As a second example of a reproducing kernel Hilbert space, consider the space of real valued functions on [0, 1] whose qth derivatives are in $\mathcal{L}_2\,[0, 1]$ with inner product

$$(f, g) = \sum_{j=0}^{q-1} \frac{f^{(j)}(0)\, g^{(j)}(0)}{(j\,!)^2} + \int_0^1 f^{(q)}(t) g^{(q)}(t)\, dt\,.$$

It can be shown that this is an R.K.H.S. with reproducing kernel

$$R_t(s) = \sum_{j=0}^{q-1} \frac{s^j t^j}{(j\,!)^2} + \int_0^{\min(s,\, t)} \frac{(s-u)_+^{q-1}\,(t-u)_+^{q-1}\, du}{((q-1)\,!)^2}$$

where $(x)_+ = x$ if $x \geqslant 0$ and $(x)_+ = 0$ if $x < 0$.

The space $H^{(1)}$ we have been discussing is the special case of this space when $q = 1$. For higher values of q, we can obtain more rapidly convergent sequences of approximations to definite integrals by finite sums with weights determined by orthogonal projection in a way similar to the previous method in $H^{(1)}$. Interpolation of a given function by piecewise polynomial functions of degree q can also be obtained by orthogonal projection in this space, using the given inner product, with higher rates of convergence than in $H^{(1)}$.

Additional examples of R.K.H. spaces with important applications can be found, for example in:

S. Haber, 'The tanh rule for numerical integration', *S.I.A.M. J. Numer. Anal.* **14**, no. 4, Sept. 1977, 668–685.

P. J. Davis, *Interpolation and approximation*, Blaisdell, N.Y., 1963, (see pp. 316–320)

R. P. Gilbert, 'Reproducing kernels for elliptic systems', *J. Approximation Theory* **15** 1975, 243–255.

There are many others as well.

In a later chapter, we will turn to applications of projections in R.K.H. spaces to approximate solutions of operator equations.

CHAPTER 10

Order relations in function spaces

In the previous introductory chapters, we have considered linear algebraic structure in function spaces as well as topological structure and relations between these two types of properties. In this chapter, we introduce a third type of structure, independent from the other two, namely *order*.

We can define another type of convergence based on order relations alone, without any linear or topological properties of the sets of functions involved.

We can define, in partially ordered sets of functions, many computationally useful things, such as intervals, monotone mappings, interval-valued mappings, lattices, and order convergence.

In linear spaces with partial ordering, we can define convex sets and convex functions.

Set inclusion is an important partial order relation in the set of all subsets of an arbitrary set.

At the beginning of the next chapter we will meet a metric which can be defined in spaces of interval valued functions.

In Chapter 14, we will discuss interval methods for operator equations which use order relations in an essential way for the design of practical computational algorithms of great generality.

The exercises in this chapter are designed to aid the reader's understanding of material in Chapter 14 particularly, and in one or two other places in subsequent chapters as well.

In addition to algebraic and topological structure, there is another important type of structure we can introduce in function spaces, namely *order.*

Definition

A relation r in a set X is called a *partial order* relation, or a *partial ordering* of X, if r is transitive.

Examples

(1) "$<$", "$\leqslant$", and "$=$" are partial order relations on the real line.

(2) In E^n, we can define the order relations

$$x < y \text{ iff } x_i < y_i,\ i = 1, 2, \ldots, n, \text{ and}$$
$$x \leqslant y \text{ iff } x_i \leqslant y_i,\ i = 1, 2, \ldots, n,$$

where $x = (x_1, x_2, \ldots, x_n)$ and $y = (y_1, y_2, \ldots, y_n)$.
This example shows why such an order relation is called a *partial* ordering. There are pairs of points $\{x, y\}$ such that we have *neither* $x \leqslant y$ *nor* $y \leqslant x$. For instance in E^2, $\{(1, 2), (2, 1)\}$ is such a pair. In E^2 we will have $x < y$ if the point y is to the right and above the point x.

(3) In the set of all real valued functions on an arbitrary set D, we can introduce the partial orderings:

$$f < g \text{ iff } f(t) < g(t) \text{ for all } t \text{ in } D; \text{ and}$$
$$f \leqslant g \text{ iff } f(t) \leqslant g(t) \text{ for all } t \text{ in } D.$$

(4) Any subset of a partially ordered set is partially ordered by the same relation.

(5) If X is any set and $S(X)$ is the set of all subsets of X, then $\subset$ and $\subseteq$ are partial orderings in $S(X)$. Note that, for U and V in $S(X)$, we have

$$U \subset V \text{ iff } x \in U \text{ implies } x \in V \text{ and } U \neq V, \text{ and}$$
$$U \subseteq V \text{ iff } x \in U \text{ implies } x \in V.$$

Note that the partial orderings $\leqslant$ and $\subseteq$ are reflexive, whereas $<$ and $\subset$ are not. The relation "=" is, of course, an equivalence relation.

Definition
Let M be a set with a *reflexive partial ordering* r. A subset of M of the form

$$A = \{x \in M : \underline{A}\ r\ x, x\ r\ \overline{A}, \text{ with } \underline{A}\ r\ \overline{A} \text{ and } \underline{A}, \overline{A} \in M\}$$

is called an *interval* in M. We denote such a set by $A = [\underline{A}, \overline{A}]$. $\underline{A}$ is called the *left endpoint* of A, and $\overline{A}$ is called the *right endpoint* of A.

We denote the set of all intervals in M by $\mathrm{II}(M)$.

If $\underline{A} = \overline{A}$, we call the interval $A = [\underline{A}, \overline{A}]$ *degenerate*. If we identify degenerate intervals with elements of M, we can view M as a subset of $\mathrm{II}(M)$. Furthermore, we have $\mathrm{II}(M) \subset S(M)$; that is, the set of all intervals in M is contained in the set of all subsets of M.

Examples

(1) For the reflexive ordering $\leqslant$ on the real line, an interval $[\underline{A}, \overline{A}]$ is an ordinary closed bounded interval of real numbers.

(2) In E^n, with the reflexive ordering $\leqslant$ defined at the beginning of this chapter, an interval is an n-dimensional rectangle of the form

$$A = [\underline{A}, \overline{A}] = ([\underline{A}_1, \overline{A}_1], \ldots, [\underline{A}_n, \overline{A}_n]).$$

We also call an interval in E^n an n-dimensional *interval vector*.

(3) In the set of all real valued functions on an arbitrary set D, the reflexive ordering $\leqslant$ defined earlier gives an interval of the form

$$F = [\underline{F}, \overline{F}] = \{f : \underline{F}(t) \leqslant f(t) \leqslant \overline{F}(t) \text{ for all } t \text{ in } D\}.$$

We can view F as a mapping $F : D \to \mathrm{II}(R)$ via $F(t) = [\underline{F}(t), \overline{F}(t)]$. This is an example of an *interval valued* mapping.

(4) Let X be an arbitrary non-empty set. Then $\subseteq$ is a reflexive partial ordering in $S(X)$ and we have intervals of the form

$$V = [\underline{V}, \overline{V}] = \{V \in S(X) : \underline{V} \subseteq V \subseteq \overline{V}\}.$$

If $X = E^2$, for instance, and if $\underline{V}$ and $\overline{V}$ are concentric discs then $[\underline{V}, \overline{V}]$ is the 'interval' of subsets of E^2 which contain the disc $\underline{V}$ and are contained in the disc $\overline{V}$, (with proper inclusion not required).

Exercise 58 If X is a set which is partially ordered by the relation r, show that, for A and B in $\mathrm{II}(X)$, we have

$$A \subseteq B \text{ iff } \underline{B}\, r\, \underline{A} \text{ and } \overline{A}\, r\, \overline{B}.$$

If $Z, Y \in S(X)$, we define $Z = Y$ to mean that Z and Y are the same subset of X. Thus $Z = Y$ iff every element of Z is an element of Y *and* vice versa. For two intervals, we have

$$Z = Y \text{ iff } \underline{Z} = \underline{Y} \text{ and } \overline{Z} = \overline{Y}.$$

Definition

A *lattice* is a set X with a reflexive partial ordering r such that for *any two* elements x and y in X there are elements w and z in X (which may depend on x and y) for which:

1. $x\, r\, w$ and $y\, r\, w$, and
   $w\, r\, u$ for any u such that $x\, r\, u$ and $y\, r\, u$; and
2. $z\, r\, x$ and $z\, r\, y$, and
   $u\, r\, z$ for any u such that $u\, r\, x$ and $u\, r\, y$.

We write $w = \sup(x, y)$ or $w = x \vee y$, and $z = \inf(x, y)$ or $z = x \wedge y$. Thus, in a lattice, every two elements have a greatest lower bound and a least upper bound. If an *arbitrary* collection of elements has a least upper bound and a greatest lower bound, then the lattice is called *complete*.

Examples

(1) Let X be an arbitrary set and consider the set $S(X)$ of subsets of X. We always consider the empty set ϕ to be an element of $S(X)$. Then $S(X)$ is a complete lattice with respect to set inclusion $\subseteq$ since, for any A and B in $S(X)$, we have

$$A \vee B = A \cup B$$
and
$$A \wedge B = A \cap B\,.$$

Furthermore, the union and intersection of an arbitrary collection of subsets of X are also in $S(X)$; thus the lattice is complete.

(2) If (X, r) is any lattice, then the set of intervals in X with the empty set adjoined is again a lattice with respect to $\subseteq$. If (X, r) is a complete lattice, then so is $\mathrm{II}(x) \cup \phi, \subseteq)$.

Exercise 59 Suppose that (X, r) is a complete lattice. Let A be a given interval in X. Denote the set of sub-intervals of A by $\mathrm{II}(A)$. Show that $(\mathrm{II}(A) \cup \phi, \subseteq)$ is a complete lattice.

Definition
Let (X, r) be a complete lattice. Suppose $x \in X$ and $\{x_n\}$ is a sequence of elements of X. We say that the sequence $\{x_n\}$ is *order convergent* to x, denoted by

$$x_n \xrightarrow{(0)} x\,,$$

iff there exist two sequences $\{y_n\}$ and $\{z_n\}$ in X such that

$$y_n \ r \ x_n \quad \text{and} \quad x_n \ r \ z_n \quad \text{for all } n = 1, 2, \ldots,$$

and
$$x = \sup\{y_n\} = \inf\{z_n\}\ .$$

Exercise 60 Show that in a complete lattice, a nested sequence of intervals with non-empty intersection is order convergent.

Definition
If $f : X \to Y$ is an arbitrary mapping from an arbitrary set X into an arbitrary set Y, we denote by $\bar{f}$ the *united extension* of f to $S(X)$, defined by

$$\bar{f} : S(X) \to S(Y)$$
$$\bar{f}(A) = \{f(x) : x \in A\} \quad \text{for } A \in S(X)\,.$$

Thus, the value of $\bar{f}$ at an element of $S(X)$ is an element of $S(Y)$. We identify the subset $\{x\}$ with its single element x in order to view $\bar{f}$ as a *set valued extension* of f. We have

$$\bar{f}\big(\{x\}\big) = \bar{f}(x) = \{f(x)\} = f(x)\,, \text{ and}$$
$$\bar{f}(A) = \bigcup_{x \in A} f(x)\,.$$

The following important property of the united extension of an arbitrary mapping is called the *subset property*:

$\forall A, B \in S(X), f : X \to Y$, we have

$$A \subseteq B \text{ implies } \bar{f}(A) \subseteq \bar{f}(B) .$$

Definitions

A mapping g from one partially ordered set (X, r_X) into another (Y, r_Y) is called *isotone* if

$$x \; r_X \; y \quad \text{implies} \quad g(x) \; r_Y \; g(y) ;$$

or *antitone* if

$$x \; r_X \; y \quad \text{implies} \quad g(y) \; r_Y \; g(x) .$$

A mapping is called *monotone* if it is either isotone or antitone.

Theorem (Knaster–Tarski, 1927). An isotone mapping of a complete lattice into itself has at least one fixed point.

Example

A closed, bounded interval $[a, b]$ on the real line is a complete lattice with respect to $\leqslant$. If a real valued function f: $[a, b] \to [a, b]$ is isotone ($x \leqslant y$ implies $f(x) \leqslant f(y)$), then f has a fixed point; that is, there is at least one x^* in $[a, b]$ such that $f(x^*) = x^*$.

Note that we do not even need continuity of f.

Exercise 61 Show by a counter-example, that an antitone real valued mapping of $[a, b]$ into $[a, b]$ need not have a fixed point.

Now let $f : X \to X$ be an *arbitrary* mapping of an *arbitrary* set X into itself. Recall that $(S(X), \subseteq)$ is a complete lattice. Consider the united extension of f,

$$\bar{f} : S(X) \to S(X) .$$

From the subset property, it follows that $\bar{f}$ is isotone with respect to the order relation $\subseteq$. From the Knaster–Tarski theorem, it follows that $\bar{f}$ has at least one fixed point in $S(X)$. Of course, this may only be the empty set. We can, however, get a very general result with important applications to numerical methods as follows.

Consider the sequence $\{X_n\}$ in $S(X)$ defined by

$$X_0 = X$$

$$X_{n+1} = \bar{f}(X_n) , \; n = 0, 1, 2, \ldots .$$

Since f maps X into itself, we have

$$X_1 = \bar{f}(X_0) = \bar{f}(X) \subseteq X = X_0 .$$

By induction, we have

$$X_{n+1} \subseteq X_n \quad \text{for all } n = 0, 1, 2, \ldots$$

since, by the subset property, we have

$$X_n \subseteq X_{n-1} \text{ implies } X_{n+1} = \bar{f}(X_n) \subseteq \bar{f}(X_{n-1}) = X_n \ .$$

Consider $Y = \bigcap_{n=0}^{\infty} X_n$, which may be empty or not. We have the following.

Theorem If $x = f(x)$ is any fixed point of f in X, then $x \in X_n$ for all $n = 0, 1, 2, \ldots$ and also $x \in Y$ and $x \in \bar{f}(Y) \subseteq Y$. Thus, X_n, Y, and $\bar{f}(Y)$ contain all the fixed points of f (if any) in X. If Y is empty or if $\bar{f}(Y)$ is empty, then there are no fixed points of f in X.

Proof If $x = f(x) \in X = X_0$, then, by induction, $x \in X_{n+1} = \bar{f}(X_n)$ for all $n = 0, 1, 2, \ldots$, and so $x \in Y$ as well. Furthermore, $Y \subseteq X_n$ for all n, so $\bar{f}(Y) \subseteq \bar{f}(X_n) = X_{n+1} \subseteq X_n$ for all n. Thus, $\bar{f}(Y) \subseteq Y$ and $x = f(x) \in \bar{f}(Y)$. By the contrapositive argument, if Y is empty or if $\bar{f}(Y)$ is empty, then there can be no x in X such that $x = f(x)$.

We will return to the application of order relations to numerical methods in a later chapter on 'interval analysis' after we have discussed some more classical approaches to the solution of operator equations. We conclude this chapter with an additional concept dealing with order relations in function spaces.

Definitions
If B is a subset of a linear space (over the real scalar field) such that the *line segment* $\overline{ab} = \{ta + (1-t)b : 0 \leqslant t \leqslant 1\}$ is contained in B whenever a and b are in B, then B is said to be *convex*.

If $f : X \to Y$ is a mapping from a linear space X (over reals) into a linear space Y with a reflexive partial ordering $\leqslant$, such that, for all a and b in X, we have

$$f(ta + (1-t)b) \leqslant tf(a) + (1-t)f(b)$$

for all $0 \leqslant t \leqslant 1$, then f is said to be a *convex function.*

Exercise 62 Show that the open and closed balls:

$$\{x : \|x\| < b\} \quad \text{and} \quad \{x : \|x\| \leqslant b\} \ ,$$

in any normed linear space, are convex sets.

Exercise 63 Show that a twice differentiable real valued function on $[a, b]$ is convex if $f''(x) \geqslant 0$ for all x in $[a, b]$.

CHAPTER 11

Operators in function spaces

In this chapter, we consider mappings (operators) from one set of functions into the same or another set of functions. In addition to normed linear spaces of functions such as Banach spaces and Hilbert spaces, we may also consider metric spaces of functions (we will presently define a metric in the set of interval valued functions), and even sets of functions which are partially ordered but without a topology. In all these, we have some notion of convergence of sequences of functions; often we will have several possible types of convergence. All these kinds of 'function spaces' will prove useful in numerical computation of approximate solutions to operator equations, as we will see in the subsequent chapters. Indeed, in this chapter we will already meet two fundamentally important methods for solving operator equations: the Neumann series, and its iterative version, Picard iteration.

In this chapter, the emphasis will be on linear operators. We will further develop the ideas and introduce additional methods for linear operator equations in the following two chapters. The six chapters following that will deal primarily with nonlinear operator equations and methods for solving them.

We begin now to put to use the tools we have studied in previous chapters to build methods for analysing and solving operator equations.

The exercises in this chapter are of particular importance. The diligent reader who works them all will be rewarded by an understanding of how to solve linear operator equations of certain kinds using the Neumann series and its iterative version. Many problems of practical importance are amenable to this approach. We will illustrate one such type of problem at the end of this chapter.

By a 'function space' we mean a set of functions together with a topology or a reflexive partial ordering (or both). Thus, we will have one or more types of convergence in a function space. We have already seen examples of linear topological spaces of functions such as Banach spaces and Hilbert spaces.

We can introduce a metric

$$d([\underline{A}, \overline{A}], [\underline{B}, \overline{B}]) = \max(|\underline{A} - \underline{B}|, |\overline{A} - \overline{B}|)$$

in the set $I(\mathbf{R})$ of intervals on the real line. (This is an example of a 'Hausdorff' metric.) The set of sub-intervals of a given interval $[\underline{A}, \bar{A}]$ is compact in this topology. Thus, we can make the set X of interval valued functions mapping $\mathrm{II}([\underline{A}, \bar{A}])$ into $\mathrm{II}(\mathbf{R})$ into a metric space with the metric

$$\mathrm{d}_X(F, G) = \sup_{A' \in \mathrm{II}([\underline{A}, \bar{A}])} \mathrm{d}(F(A'), G(A')) .$$

On the other hand, we might consider such function spaces as real valued functions on a set D with the reflexive partial ordering $f \leqslant g$ iff $f(x) \leqslant g(x)$ for all x in D or set valued functions mapping $S(D_1)$ into $S(D_2)$ with $F \subseteq G$ iff $F(A) \subseteq G(A)$ for all $A \in S(D_1)$.

Definition

An *operator* is a mapping from one function space into another (or the same) function space. If X and Y are linear spaces (over the reals), then a mapping $L : X \to Y$ is called a *linear* operator if

$$L(f+g) = L(f) + L(g)$$

and

$$L(af) = aL(f)$$

for all f, g in X and all real a. Notice that we will always have $L(0) = 0$ for a linear operator.

An operator which is not linear is called *nonlinear.*

Examples

(1) For $X = Y = E^n$ (which can be viewed as the linear space of real valued functions on the first n positive integers), a linear transformation, represented by an n-by-n matrix, is a linear operator.

(2) For $X = C[0, 1]$ and $Y = \mathbf{R}$, a linear functional is a linear operator, from X into Y.

(3) For $X = Y = C[0, 1]$ and K continuous on $[0, 1] \times [0, 1]$,

$$L(f)(t) = \int_0^1 K(t, t') f(t') \, \mathrm{d}t'$$

defines a linear *integral* operator, mapping X into itself.

(4) For $X = C^2[0, 1]$, twice continuously differentiable real valued functions on $[0, 1]$, and $Y = C[0, 1]$,

$$L(f)(t) = a(t) f''(t) + b(t) f'(t) + c(t) f(t)$$

defines a linear *differential* operator L from X into Y, if a, b and c are continuous.

(5) The Laplace operator

$$\Delta = \frac{\partial^2}{\partial x^2} + \frac{\partial^2}{\partial y^2} + \frac{\partial^2}{\partial z^2}$$

is a linear differential operator on the space of real valued functions with continuous second partial derivatives on a region in E^3 into continuous functions on that region.

(6) The Fourier transform

$$F(f) = \int_{-\infty}^{\infty} e^{2\pi i s t} f(t)\,dt\,, \quad -\infty < s < \infty\,,$$

is a linear integral operator on a certain space of complex valued functions.

(7) If X is the linear space of continuously differentiable n-dimensional real vector valued functions on the real line, then $L(x)\,(t) = x'(t) - A(t)x(t)$, with $A(t)$ an n-by-n matrix with continuous elements, defines a linear differential operator from X into continuous vector valued functions.

(8) $P : C[0, 1] \to C[0, 1]$ is a *nonlinear* operator, for example, if $P(f) = g \neq 0$ for all f in $C[0, 1]$ or if $P(f)\,(t) = e^{f(t)}$.

Numerous other examples will occur in the chapters to follow.

Definitions

If $L : X \to Y$ is a linear operator (X and Y are linear spaces), then $N(L) = \{x : x \in X,\ L(x) = 0\}$ is called the *null space* of L. It is the set of points mapped into the zero vector, or 'annihilated', by L. If $y \in Y$, then $L(x) = y$ is called a *linear operator equation*. We often write this as $Lx = y$.

Exercise 64 Show that the null space of a linear operator is a linear manifold.

The linear operator L is called *non-singular* iff $N(L)$ consists only of the zero vector in X otherwise, it is called *singular*.

Exercise 65 Show that if a linear operator equation $Lx = y$ has a solution x, then the solution is unique if L is non-singular.

Exercise 66 Show that, if a linear operator $L : X \to Y$ is non-singular and maps X *onto* Y, then the equation $Lx = y$ has a unique solution for every y in Y.

Exercise 67 Let $C^1(\mathbf{R})$ be the linear space of continuously differentiable real valued functions on the real line. Consider the linear space of functions

$$Y = \{y = Lx : x \in C^1(\mathbf{R}),\ (Lx)\,(t) = x'(t) - a(t)x(t)\}$$

where a is a given element of $C(\mathbf{R})$. Thus $L : C^1(\mathbf{R}) \to Y$. Find the null space of L. Suppose the linear operator equation $Lx = b$ has a solution x^* for a certain b in Y. Find the set of *all* solutions to $Lx = b$.

Definition

Let X and Y be normed linear spaces. A linear operator $L : X \to Y$ is *continuous* at $x \in X$ iff, $\forall \{X_n\} \subseteq X$,

$$\lim_{n\to\infty} \|x_n - x\|_X = 0 \text{ implies } \lim_{n\to\infty} \|Lx_n - Lx\|_Y = 0.$$

Exercise 68 Show that if a linear operator $L : X \to Y$ is continuous at $x = 0$, then it is continuous at every $x \in X$.

Definitions

A linear operator $L : X \to Y$ is called *bounded* (X and Y are normed linear spaces) iff $\exists$ $b \geqslant 0$ such that $\|Lx\|_Y \leqslant b\|x\|_X$ for all x in X.

The *norm* $\|L\|$ of a bounded linear operator $L : X \to Y$ is defined as

$$\|L\| = \sup_{\|x\|_X=1} \|Lx\|_Y .$$

Exercise 69 Show that a linear operator is continuous iff it is bounded.

Exercise 70 Let $A : X \to X$ be a bounded linear operator on a Banach space X, such that $\|A\| < 1$. Let $I : X \to X$ be the *identity* operator $Ix = x$ for all x in X. Consider the linear operator

$$L = I - A \text{ defined by } Lx = x - Ax, \ \forall x \in X.$$

Show that L is one-one, that L^{-1} exists and can be represented by the *Neumann series*

$$L^{-1} = I + A + A^2 + A^3 + \ldots\ldots \ [A^n = A \circ (A^{n-1})] .$$

Show, further, that $Lx = y$ has a unique solution for every y in X, and that the sequence $\{x_n\}$ defined by

$$x_{n+1} = y + Ax_n , \quad n = 0, 1, 2, \ldots ,$$

converges to $x = L^{-1}y$ from any starting point x_0 in X. Derive an error bound on $\|x - x_n\|$.

Exercise 71 Using the results of the previous exercise, show that the two-point boundary value problem

$$x''(t) + a(t)\, x(t) = b(t)$$

$$x(0) = x(1) = 0$$

has a unique solution for every continuous b, provided a is continuous and

$$\max_{t \in [0,1]} |a(t)| < c$$

for a certain positive real number c. Find a numerical value we can use for c in this result. (Hint: rewrite the boundary value problem as an integral equation

$$x(t) = y(t) = \int_0^1 K(t, s)a(s)x(s)\,\mathrm{d}s$$

where
$$K(t, s) = \begin{cases} t(1-s), & 0 \leqslant t \leqslant s \leqslant 1 \\ s(1-t), & 0 \leqslant s \leqslant t \leqslant 1 \end{cases}$$

and
$$y(t) = -\int_0^1 K(t, s)b(s)\,\mathrm{d}s \text{ .)}$$

Exercise 72 For the problem in the previous exercise, put $a(t) = 1$ for all t and put $b(t) = t$. Set $x_0(t) = 0$ for all t and find $x_1(t)$ defined in Exercise 70 explicitly. Also find a numerical bound on

$$\|x - x_1\| = \max_{t \in [0,1]} |x(t) - x_1(t)|$$

where x is the *unknown* solution.

ADJOINT OPERATORS

Let $L : X \to Y$ be a bounded linear operator, X and Y Banach spaces, and let X^* and Y^* be the conjugate spaces of bounded linear functionals on X and Y. The *adjoint* operator L^* (of the operator L) is defined by $L^* : Y^* \to X^*$ where $(L^*g)(x) = g(Lx)$ for $x \in X$ and $g \in Y^*$. Thus, $L^*g \in X^*$ is a bounded linear functional on X. The adjoint L^* is also a bounded linear operator and $\|L^*\| = \|L\|$.

Exercise 73 Let H be a Hilbert space and $L : H \to H$ a bounded linear operator. Show that L^* is the adjoint of L if

$$(x, Ly) = (L^*x, y) \text{ for all } x, y \text{ in } H.$$

If $L^* = L$, then L is called *self-adjoint*.

It can be shown that if L and L^* are linear operators in a Hilbert space for which $(x, Ly) = (L^*x, y)$ for all x, y in H, then L is bounded and L^* is the adjoint of L. A special case (Hellinger and Toeplitz, 1910) is the following: if $L : H \to H$ admits a matrix representation with respect to some orthogonal basis in a separable Hilbert space H, then L is bounded. For the converse, we have the following.

Theorem A bounded linear operator L on a separable Hilbert space H, $L : H \to H$, can be represented by a matrix.

Proof Let $\{e_k\}$ be an orthonormal basis for H. Then any $x \in H$ can be written as

$$x = \sum_{k=1}^{\infty} x_k e_k \quad \text{with } x_k = (x, e_k).$$

Furthermore, we have

$$Lx = \sum_{k=1}^{\infty} x_k (Le_k) = \sum_{k=1}^{\infty} x_k \left(\sum_{i=1}^{\infty} L_{ik} e_i \right)$$

where $L_{ik} = (Le_k, e_i)$. Thus

$$Lx = \sum_{i=1}^{\infty} \left(\sum_{k=1}^{\infty} L_{ik} x_k \right) e_i$$

can be represented in the usual way (as for linear operators in E^n) with matrix notation:

$$\begin{pmatrix} L_{11} & L_{12} & L_{13} & \cdots \\ L_{21} & L_{22} & L_{23} & \cdots \\ L_{31} & \cdots & & \\ \cdots & & & \\ \cdots & & & \end{pmatrix} \begin{pmatrix} x_1 \\ x_2 \\ x_3 \\ \cdot \\ \cdot \end{pmatrix} = \begin{pmatrix} (Lx)_1 \\ (Lx)_2 \\ (Lx)_3 \\ \cdot \\ \cdot \end{pmatrix}$$

where $$Lx = \sum_{i=1}^{\infty} (Lx)_i e_i.$$

The convergence of the various series involved follows from the continuity of the bounded linear operator L. In particular, for all k we have

$$\sum_{i=1}^{\infty} |L_{ik}|^2 < \infty.$$

Definition
We can make the set of all bounded linear operators mapping a Hilbert space H into itself into a linear space $L(H, H)$ with

$$(aL_1 + bL_2)x = a(L_1 x) + b(L_2 x).$$

Exercise 74 Let A_n be a sequence of bounded linear operators in $L(H, H)$. Suppose that $A \in L(H, H)$. Show that the following hierarchy of types of convergence holds:

uniform convergence, $A_n \Rightarrow A$ iff $\|A_n - A\| \to 0$,

implies

strong convergence, $A_n \to A$ iff $A_n x \to Ax, \ \forall x \in H$,

implies

weak convergence, $A_n \xrightarrow{w} A$ iff $(A_n x, h) \to (Ax, h), \ \forall x, h \in H$.

Important *unbounded* operators are usually *densely defined*, that is they are defined on dense subsets of a given space. For example the differential operator $\mathrm{d}/\mathrm{d}t$ is densely defined in $C[0, 1]$ by

$$(\mathrm{d}/\mathrm{d}t)x(t) = \lim_{h \to 0} \frac{x(t+h) - x(t)}{h}$$

for differentiable functions in $C[0, 1]$. This operator is an unbounded linear operator. Since polynomials are dense in $C[0, 1]$ by the Weierstrass approximation theorem, so are differentiable functions (which include the polynomials).

Note that the recursive generation of a sequence $\{x_n\}$ which converges to the unique solution of a linear operator equation $(I - A)x = y$ given in Exercise 70, with $\|A\| < 1$, provides an iterative method for such problems. In the special case that $X = E^n$ and A is an n-by-n matrix, it is called the *Jacobi iteration method*. When applied to an integral equation such as in Exercise 71, it is usually called *Picard iteration*.

We will now illustrate an application of the Neumann series and Picard iteration to a class of initial value problems of a type which occurs frequently in practical applied mathematics.

We consider a system of linear first order ordinary differential equations, which can be written in vector form as

$$x'(t) = h(t) + M(t)x(t)$$

with given initial conditions $x(t_0) = a$.

We will suppose that h is a vector of real valued functions which are continuous on the real line and that M is a matrix of real valued functions which are continuous on the real line. Thus, for each t, suppose that $h(t)$ is an n-dimensional vector and $M(t)$ is an n-by-n matrix. The initial point $x(t_0) = a$ is also an n-dimensional vector. Such a problem includes, as a special case, when $h(t)$ is identically zero and $M(t)$ is a matrix of constants, a linear differential system with constant coefficients such as occurs frequently in linear stability analysis for vibrating structures and oscillating electrical networks. The following method can be used for a detailed analysis of transient behaviour.

We can rewrite such a problem as an integral equation

$$x(t) = y(t) + \int_{t_0}^{t} M(s)x(s)\,ds\,,$$

where $$y(t) = a + \int_{t_0}^{t} h(s)\,ds\,.$$

This form, which is more suitable for application of the Neumann series method, is obtained from the differential equation and initial conditions by integrating both sides of the differential equation and applying the initial conditions.

If we can obtain a solution for t in $[t_0, t_1]$, then we can continue the solution beyond t_1 conveniently by making use of the so-called 'semi-group property' for initial value problems, that is,

$$x(t) = y(t; t_1) + \int_{t_1}^{t} M(s)x(s)\,ds\,,$$

where $$y(t; t_1) = x(t_1) + \int_{t_1}^{t} h(s)\,ds\,.$$

In component form, the integral equation can be written

$$x_i(t) = y_i(t; t_1) + \int_{t_1}^{t} \sum_{j=1}^{n} M_{ij}(s)x_j(s)\,ds,$$

where $$y_i(t; t_1) = x_i(t_1) + \int_{t_1}^{t} h_i(s)\,ds\,,$$

and $i = 1, 2, \ldots, n$.

Next, we can consider the first integral equation we wrote down above as having the form $x = y + Ax$ or $(I - A)x = y$, where A is the linear operator defined by

$$[Ax](t) = \int_{t_0}^{t} M(s)x(s)\,ds\,.$$

We can regard A as a bounded linear operator on vector valued functions with components in $C[t_0, t_1]$, with $t_1 > t_0$ to be determined. Using the usual uniform norm in $C[t_0, t_1]$, we find that

$$\|A\| \leqslant \int_{t_0}^{t_1} \|M(s)\|\,ds \leqslant (t_1 - t_0)\,\|M\|,$$

with $\|x\| = \max_i \|x_i\|$ and $\|M\| = \max \|M(s)\|$ for s in $[t_0, t_1]$, where $\|x_i\| = \max |x_i(s)|$ for s in $[t_0, t_1]$ and $\|M(s)\| = \max_i \sum_j |M_{ij}(s)|$. Note that we could also write $\|x\| = \max_{t_0 \leqslant s \leqslant t_1} (\max_i |x_i(s)|)$. Thus, we will have

$$\|A\| < 1 \text{ if } (t_1 - t_0)\|M\| < 1.$$

Let $[a, b]$ be an arbitrary interval on the real line, then $\|M\|$, with the norm taken over $[t_0, t_1]$ for t_0 and t_1 in $[a, b]$ is no greater than $\|M\|_{[a,b]} = \max \|M(s)\|$ for s in $[a, b]$. It follows that we can continue the solution, using the Neumann series method, up to an arbitrary value b, from a given t_0, since we can go in steps of size $1/\|M\|_{[a,b]}$, using the semi-group property, while keeping $\|A\| < 1$.

A simple numerical example will help to illustrate the application of the method. Consider the following scalar initial value problem:

$$x'(t) = 1 + tx(t) \text{ with } x(0) = 1.$$

In this example, we have $y(t) = 1 + t$, $M(t) = t$, and

$$[Ax](t) = \int_0^t sx(s)\,ds.$$

It follows that, in $C[0, t_1]$, we have $\|M\| = t_1$ for $t_1 > 0$, and $\|A\| \leqslant t_1^2/2 = \int_0^{t_1} |M(s)|\,ds = \int_0^{t_1} s\,ds$. Thus, $\|A\| < 1$ if $t_1 < \sqrt{2}$. Thus, for t in $[0, t_1]$ with $t_1 < \sqrt{2}$, we can put $x_0(t) = 0$, and the iterations defined by

$$x_{n+1}(t) = y(t) + \int_0^t sx_n(s)\,ds, \quad n = 0, 1, 2, \ldots,$$

will converge uniformly to the solution in $[0, t_1]$. For the first few iterations, we find

$$x_1(t) = 1 + t,$$
$$x_2(t) = 1 + t + t^2/2 + t^3/3,$$
$$x_3(t) = 1 + t + t^2/2 + t^3/3 + t^4/8 + t^5/15.$$

Since the solution satisfies

$$x = (I - A)^{-1} y = (I + A + A^2 + \ldots)y,$$

and $$x_n = (I + A + A^2 + \ldots + A^n)y,$$

we have $$x - x_n = A^{n+1}(I + A + A^2 + \ldots)y.$$

It follows that

$$\|x - x_n\| \leqslant \|A\|^{n+1} \|y\|/(1 - \|A\|).$$

In particular, for $t_1 = 1$, we have $\|A\| \leqslant 1/2$ and so we obtain

$$\|x - x_n\| \leqslant (0.5)10^{-6}, \quad \text{for } n \geqslant 22, \quad \text{on } [0, 1].$$

If we choose a smaller t_1, say $t_1 = 0.1$, then we will have faster convergence to a given accuracy; in fact, for this t_1, we will have

$$\|x - x_n\| \leqslant (0.14)10^{-6}, \quad \text{for } n \geqslant 2, \quad \text{on } [0, 0.1].$$

Thus, the iterative method can produce an approximation of arbitrary accuracy at any point on the real line in this example. Of course, in order to go beyond some $t_1 < \sqrt{2}$, we would have to make use of the semi-group property and start a new initial value problem with a solution value found in a previous interval.

To use such a method in practice, one would want to first decide on efficient choices of steps in the continuation process, so that not too many iterations would be needed in order to obtain the desired accuracy in each interval. We will not pursue this interesting question here since it would lead to far afield from our main subject.

In the simple numerical example just discussed, we can, of course, represent the exact solution in the form

$$x(t) = \exp(t^2/2)\left\{1 + \int_0^t \exp(-s^2/2)\,\mathrm{d}s\right\}.$$

The same method, however, can be applied to much more complicated examples, involving large systems of linear differential equations, for which no such easy closed form solution exists. For a more formidable example, we would, of course, program a computer to carry out the necessary computations.

CHAPTER 12

Completely continuous (compact) operators

In this chapter, we introduce a large class of linear operators called *completely continuous* (or *compact*) operators. They frequently occur in linear operator equations. It is often the case that the Neumann series method cannot be applied to such operator equations. Fortunately, there are other methods available for such equations. We discuss two such methods in this chapter: eigenfunction expansions and the Galerkin method. Further discussion of these and other methods for linear operator equations can be found in the next chapter.

Two important types of completely continuous operators are Hilbert–Schmidt integral operators and projections into finite dimensional subspaces. We will concentrate on these in this chapter.

Completely continuous linear operators constitute a generalization of linear operators in finite dimensional vector spaces. Both can be represented by matrices, in the case of inner product spaces. In addition, we have the so-called 'Fredholm alternative' for such operators, which is analogous to the following situation for finite dimensional linear algebraic systems:

for $\qquad (I-A)x = b\,,$

only one of the following two alternatives is possible

1) there is a unique solution vector x for every b and $(I-A)$ has an inverse; or
2) $(I-A)x = 0$ has a non-trivial solution and $(I-A)$ is singular and $+1$ is an eigenvalue of the matrix A.

Let L be a bounded linear operator $L: H \rightarrow H$ with H a separable Hilbert space. Recall that if $\|L\| < 1$, then the linear operator equation $(I-L)x = y$ has a unique solution for every y in H.

Another important class of bounded linear operators, with $\|L\|$ not necessarily less than 1, for which $(I-L)x = y$ has a unique solution, is the class of completely continuous operators (also called compact operators). We proceed to outline the theory of these operators and discuss computational methods based on the theory, including the method of eigenfunction expansions and Galerkin's method.

Definition

Let H be a separable Hilbert space. A linear operator $L : H \to H$ is *completely continuous* (*compact*) if L maps weakly convergent sequences into strongly convergent sequences (equivalently, if L maps bounded sets into relatively compact sets – in the sense of strong convergence).

Exercise 75 Show that a completely continuous (c.c.) operator is necessarily bounded. Show that the identity mapping on an infinite dimensional separable Hilbert space is *not* c.c.

Examples

(1) It can be shown that if $L : H \to H$, H a separable Hilbert space, has a matrix representation (L_{ij}) such that

$$\sum_{i,j} |L_{ij}|^2 < \infty ,$$

then L is c.c.

(2) Consider $L : \mathcal{L}_2[a, b] \to \mathcal{L}_2[a, b]$ defined by

$$(Lx)(t) = \int_a^b K(t, s)x(s)\,\mathrm{d}s \ , \ t \in [a, b] .$$

The function K is called the *kernel* of the linear integral operator L.

If $$\int_a^b \int_a^b |K(t, s)|^2 \,\mathrm{d}s\,\mathrm{d}t < \infty ,$$

then K is called a *Hilbert–Schmidt kernel* and L is called a *Hilbert–Schmidt operator*. Such an operator L is bounded; in fact

$$\|L\| < \left\{ \int_a^b \int_a^b |K(t, s)|^2 \,\mathrm{d}s\,\mathrm{d}t \right\}^{1/2} .$$

It follows that a Hilbert–Schmidt (H.–S.) operator has a matrix representation with respect to any given orthonormal basis in $\mathcal{L}_2[a, b]$. It can be shown that

$$\sum_{i,j} |L_{ij}|^2 = \int_a^b \int_a^b |K(t, s)|^2 \,\mathrm{d}s\,\mathrm{d}t ,$$

so $$\left(\sum_{i,j} |L_{ij}|^2 \right)^{1/2} < \infty .$$

It follows that an H.–S. operator is completely continuous.

(3) Let G be a non-trivial subspace of a separable Hilbert space H. Thus $g \neq 0$ for some $g \in G$. Write $H = G \oplus F$; that is, $h \in H$ implies that $h = g + f$ with $g \in G$ and $f \in F$ and $(g, f) = 0$. Consider the particular *projection operator*

$$\Pi_G : H \to G \subseteq H$$

defined by $\Pi_G\, h = g$, where $h = g + f$, $g \in G$, $f \in F$, and $(g, f) = 0$. The following are properties of Π_G:

1) Π_G is a bounded linear operator and $\|\Pi_G\| = 1$.
2) Π_G is *idempotent*: $\Pi_G\,(\Pi_G\, h) = \Pi_G\, h, \forall\, h \in H$.
3) Π_G is self-adjoint, $\Pi_G^* = \Pi_G$.
4) If G_1 and G_2 are subspaces of H, then

$$G_2 \subseteq G_1 \text{ iff } \|\Pi_{G_2}\, h\| \leqslant \|\Pi_{G_1}\, h\|, \forall\, h \in H.$$

5) If G is finite dimensional, then Π_G is completely continuous.

Exercise 76 Prove (1) – (5) above.

Exercise 77 We can define the *product* of two operators as the composition of the operators; thus AB is the operator defined by $(AB)x = A(Bx)$. Show that if A is c. c. and B is bounded, then AB and BA are c. c. Thus c. c. operators form an ideal in the ring of bounded linear operators.

Exercise 78 If A is c. c. on an infinite dimensional separable Hilbert space, then it does not have an inverse defined on the whole space. Why? It may have an inverse on a dense subset of the space.

Next, we come to the *spectral theory* of c. c. operators.

Definition
Let $A : H \to H$ be a completely continuous linear operator on a separable Hilbert space H. If

$$Ax = \lambda x \text{ for some } x \neq 0 \text{ in } H, \text{ then}$$

λ is called an *eigenvalue* of A and x an *eigenvector*.

It can be shown† that: a c. c. operator $A : H \to H$ has only a finite number of eigenvalues whose moduli exceed any given positive real number; if A is self-adjoint, the eigenvalues of A are all real; the only possible limit point of the eigenvalues of A is 0; corresponding to each nonzero eigenvalue of A, there are at most a finite number of linearly independent eigenvectors; any nonzero, c. c.,

---

† See N. I. Akhieser and I. M. Glazman, *Theory of linear operators in Hilbert spaces*, Vol. I, Ungar, N.Y., 1966.

self-adjoint linear operator A has at least one nonzero eigenvalue and a finite or infinite orthonormal sequence of eigenvectors $e_1, e_2, \ldots$ belonging to the set of its nonzero eigenvalues, $|\lambda_1| \geqslant |\lambda_2| \geqslant |\lambda_3| \geqslant \ldots\ldots$, such that, for each vector y in the range of A ($Ax = y$ for some x in H), y can be represented as

$$y = \sum_{k=1}^{\infty} (y, e_k)e_k$$

with
$$\lim_{N\to\infty} \|y - \sum_{k=1}^{N} (y, e_k)e_k\| = 0.$$

If A is c. c., then $I - A$ has an inverse $(I - A)^{-1}$ which is a bounded linear operator on H iff $+ 1$ is *not* an eigenvalue of A (iff $(I - A)x = 0$ implies $x = 0$).

If A is c. c. and self-adjoint and if $+ 1$ is not an eigenvalue of A, then there is a sequence of nonzero eigenvalues $\{\lambda_k\}$ and a sequence of orthonormal eigenvectors (also called *eigenfunctions* of A, if elements of H are functions), $\{e_k\}$, with $Ae_k = \lambda_k e_k$ and the linear operator equation $(I - A)x = y$ has a unique solution x for every y in the range of A. The solution x can be represented by

$$x = \sum_{k=1}^{\infty} c_k e_k ,$$

and the coefficients c_k can be found as follows.

We put
$$(I - A)x = (I - A) \sum_{k=1}^{\infty} c_k e_k = \sum_{k=1}^{\infty} (y, e_k)e_k ;$$

thus,
$$\sum_{k=1}^{\infty} c_k(1 - \lambda_k)e_k = \sum_{k=1}^{\infty} (y, e_k)e_k .$$

Equating coefficients of e_k on both sides, we obtain

$$c_k(1 - \lambda_k) = (y, e_k) ,$$

or
$$c_k = (y, e_k)/(1 - \lambda_k) .$$

Thus, $(I - A)x = y$ has the solution

$$x = \sum_{k=1}^{\infty} [(y, e_k)/(1 - \lambda_k)]\, e_k .$$

Exercise 79 Show that a Hilbert–Schmidt operator is self-adjoint if its kernel is symmetric ($K(t, s) = K(s, t)$). Write the boundary value problem,

$$x''(t) + x(t) = e^t , \quad x(0) = x(1) = 0 ,$$

as an integral equationwith a H.–S. operator A in the form $(I - A)x = y$.

Find the eigenvalues and eigenfunctions of A. Solve the equation for x in an eigenfunction expansion.

If we can find the first N nonzero eigenvalues of A, $|\lambda_1| \geqslant |\lambda_2| \geqslant \ldots \geqslant |\lambda_N|$, and corresponding linearly independent eigenvectors $e_1, e_2, \ldots, e_N$, then we can approximate the solution of $(I - A)x = y$ by *projecting* the eigenfunction expansion into the subspace G_N spanned by $\{e_1, e_2, \ldots, e_N\}$

$$x \approx \sum_{k=1}^{N} [(y, e_k)/(1 - \lambda_k)] e_k = \Pi_{G_N} \left(\sum_{k=1}^{\infty} [(y, e_k)/(1 - \lambda_k)] e_k \right) .$$

Thus, the projection Π_{G_N} produces just the first N terms of the series.

Exercise 80 Show that $\|x - \Pi_{G_N} x\| \to 0$ as $N \to \infty$.

There is another, more general, approach to obtaining approximate solutions using orthonormal basis functions, known as *Galerkin's method*, which we describe now.

Let $A : H \to H$ be c. c., H a separable Hilbert space, and consider the linear operator equation $(I - A)x = y, y \in H$.

Let G_n be an n-dimensional subspace of H and consider the related operator equation

$$(I - \Pi_{G_n} A)x_n = \Pi_{G_n} y$$

where Π_{G_n} is the projection of an element of H into G_n, and $(\Pi_{G_n} A)x$ denotes the composition $\Pi_{G_n}(Ax)$.

Suppose that $\{e_1, e_2, \ldots\}$ is an orthonormal basis for H. Let G_n be the subspace spanned by $\{e_1, e_2, \ldots, e_n\}$. Then

$$\Pi_{G_n} h = \sum_{k=1}^{n} (h, e_k) e_k \quad \text{for all } h \in H.$$

Put $x_n = \sum_{k=1}^{n} c_k^{(n)} e_k$. We seek the coefficients $c_1^{(n)}, \ldots, c_n^{(n)}$ such that

$$(I - \Pi_{G_n} A) \sum_{k=1}^{n} c_k^{(n)} e_k = \sum_{k=1}^{n} (y, e_k) e_k .$$

Now

$$\Pi_{G_n} A \sum_{k=1}^{n} c_k^{(n)} e_k = \sum_{R=1}^{n} c_k^{(n)} \Pi_{G_n} (Ae_k)$$

$$= \sum_{k=1}^{n} \left(c_k^{(n)} \sum_{j=1}^{n} (Ae_k, e_j) e_j \right) .$$

If we now form the inner products of e_i with both sides of the operator equation $(I - \prod_{G_n} A)x_n = \prod_{G_n} y$ and use the expression derived above, we obtain the following n-dimensional linear algebraic system of equations to solve for $c_1^{(n)}, \ldots, c_n^{(n)}$:

$$c_i^{(n)} - \sum_{k=1}^{n} c_k^{(n)} (Ae_k, e_i) = (y, e_i), \; i = 1, 2, \ldots, n.$$

If we can solve this system for $c_1^{(n)}, \ldots, c_n^{(n)}$, then we obtain an approximation

$$x_n = \sum_{i=1}^{n} c_i^{(n)} e_i$$

to the solution of $(I - A)x = y$.

Variations of this method are variously known as the Rayleigh–Ritz method, the method of moments, or simply variational methods. (See, for example, L. M. Delves and J. Walsh, *Numerical solution of integral equations*, Oxford University Press, 1974; I. Babuska, M. Prager, and E. Vitasek, *Numerical processes in differential equations*, Wiley, 1966. These references contain discussions of error estimates for various Galerkin-type methods in Hilbert spaces.)

If we call $d_n = x - x_n$, and if $I - A$ is positive definite and bounded below,

$$(u, (I - A)u) \geqslant b(u, u) \text{ for all } u \text{ in } H, \text{ for some } b > 0,$$

then
$$\|d_n\|^2 = (d_n, d_n) \leqslant (1/b)(d_n, (I - A)d_n) \leqslant (1/b)\|d_n\| \, \|(I - A)d_n\|;$$

thus
$$\|d_n\| = \|x - x_n\| \leqslant (1/b)\|(I - A)x_n - y\|.$$

If we can find an appropriate number $b > 0$, then we can evaluate an upper bound on $\|x - x_n\|$ after having found x_n. In this way we can sometimes find an *a posteriori* bound on $\|x - x_n\|$. For the number b, we can take $1 - \|A\|$ in the case that $\|A\| < 1$, since we will then have

$$\begin{aligned}(1 - \|A\|)(u, u) &= (u, u) - \|A\|(u, u) \\ &\leqslant (u, u) - |(u, Au)| \\ &\leqslant (u, u) - (u, Au) \\ &\leqslant (u, u - Au) \text{ for all } u \text{ in } H.\end{aligned}$$

Some of the results just discussed in Hilbert spaces can be extended to Banach spaces.

Definition
A linear operator A which maps a Banach space into itself is called *completely continuous* iff it maps an arbitrary bounded set into a relatively compact set.

Example
In the Banach space $C[a, b]$, an integral operator L of the form

$$L(x)(t) = \int_a^b K(t, s)x(s)\,ds$$

is completely continuous if $K(t, s)$ is bounded for $s, t \in [a, b]$ and K is continuous on $[a, b] \times [a, b]$ except possibly on a finite number of continuous curves. (See A. N. Kolmogorov and S. V. Fomin, *Elements of the theory of functions and functional analysis*, Vol. 1, Graylock Press, 1957, pp 112–113).

If $\{A_n\}$ is a sequence of c. c. operators on a Banach space B and A is a bounded linear operator on B, $A_n : B \to B$ and $A : B \to B$, such that $\|A_n - A\| \to 0$, then A is also c. c.

If $A : B \to B$ is c. c. and $C : B \to B$ is a bounded linear operator, then AC and CA are c. c.

A spectral theory analogous to that for Hilbert spaces exists for c. c. operators in Banach spaces.

The following is known as *the Fredholm alternative.*

Theorem Let $A : B \to B$ be a completely continuous linear operator, B a Banach space, and consider the linear operator equation

$$(I - A)x = y, \quad y \in B.$$

Only one of the two following alternatives is possible:

1) the equation has a unique solution for every y in B; and $(I - A)$ has an inverse; or
2) $(I - A)x = 0$ has a nonzero solution; and $(I - A)$ is singular and 1 is an eigenvalue of A.

For a discussion of Galerkin's method in Banach spaces with application to the solution of Fredholm integral equations of the second kind, see: Y. Ikebe, *S.I.A.M. Review* **14** no. 3, 1972, pp. 465–491.

CHAPTER 13

Approximation methods for linear operator equations

In this chapter, we will discuss a number of methods for the approximate solution of linear operator equations. In doing so, we will illustrate applications of the concepts and techniques of functional analysis which have been discussed in the preceding chapters. We will not attempt the most general treatment of each method; but rather, we will illustrate each method on a specific type of operator equation.

The methods will, for the most part, fall into two categories: (1) *finite basis methods*, and (2) *finite difference methods*.

In later chapters, we will consider methods for nonlinear operator equations.

Suppose we wish to solve a linear operator equation $Lx = y$ where $L : C(0, 1] \to C[0, 1]$ is a linear operator for which L^{-1} exists. Then a unique solution, $x = L^{-1}y$ exists for any y in $C[0, 1]$. It may be, however, that a representation of L^{-1} involves an infinite series. If we wish to obtain numerical values of the solution, we will be limited to a finite number of arithmetic operations in practical applications. Therefore, we are interested in methods which *approximate* the exact solution, together with methods for estimating or bounding the error in such approximations.

In a *finite basis method*, we choose a finite set of basis elements of $C[0, 1]$, say $f_1, f_2, \ldots, f_n$, and give a procedure for finding coefficients $c_1, c_2, \ldots, c_n$ to obtain an approximate solution of the form

$$x_n = \sum_{k=1}^{n} c_k f_k \ .$$

Note that such an approximation provides a continuous approximation to the solution, since $x_n \in C[0, 1]$.

In contrast to this type of method, a *finite difference method* provides only a 'discrete' approximation consisting of approximations to $x(t)$ on a finite set of values of t, say $t_1, t_2, \ldots, t_n$ in $[0, 1]$. Of course, with finite difference methods, we can interpolate the discrete approximate solution to obtain a continuous approximation. In fact, as we will see, this is one way to obtain bounds on the error in the discrete approximation using the methods of functional analysis.

In the previous two chapters, we have already discussed *three methods* for the solution of linear operator equations $Lx = y$, when L is of the form $L = I - A$, under certain conditions on A. The first of these was *an iterative method based on the Neumann series* when $\|A\| < 1$. The second and third were *the method of eigenfunction expansions* and *Galerkin's method*. These latter two can be viewed as finite basis methods (when we seek only the first N terms of the eigenfunction expansion); for the second and third methods, A is assumed to be a c. c. operator which does not have 1 as an eigenvalue.

The classical method of combining Fourier series expansions with the method of separation of variables for certain types of boundary value problems for parabolic type partial differential equations illustrates an application of the method of eigenfunction expansions. Consider the equation (heat conduction, diffusion, etc.)

$$\frac{\partial \bar{u}}{\partial \bar{t}} = \bar{k}\,\frac{\partial^2 \bar{u}}{\partial \bar{x}^2}\ ,\ 0 < \bar{t}\ ,\ 0 < \bar{x} < L,$$

with boundary conditions, for constants $\bar{u}_0$ and $\bar{u}_L$,

$$\bar{u}(0, \bar{t}) = \bar{u}_0\ ,\ \bar{t} > 0,$$

$$\bar{u}(L, \bar{t}) = \bar{u}_L\ ,\ \bar{t} > 0,$$

and initial conditions

$$\bar{u}(\bar{x}, 0) = g(\bar{x})\ ,\ 0 \leqslant \bar{x} < L\ .$$

We can make an easy change of variables so that the problem takes the form

$$u_t = u_{xx}\ ,\ 0 < t,\ 0 < x < 1,$$

with $\quad u(0, t) = u(1, t) = 0,\ t > 0$

and $\quad u(x, 0) = f(x) = g(Lx) - \bar{u}_0 - (\bar{u}_L - \bar{u}_0)x,\ 0 \leqslant x \leqslant 1.$

If we can find $u(x, t)$, or an approximation to it, then we can construct

$$\bar{u}(\bar{x}, \bar{t}) = \bar{u}_0 + (\bar{u}_L - \bar{u}_0)x + u(x, t)$$

with $\quad \bar{t} = L^2 t/\bar{k}\ \text{ and }\ \bar{x} = Lx.$

In the method of separation of variables, we put

$$u(x, t) = a(x)b(t)\ .$$

Substituting this into the differential equation, we have

$$a(x)b'(t) = a''(x)b(t)\ ,\ \text{or}$$

$$a''(x)/a(x) = b'(t)/b(t),\ \text{for } a(x) \neq 0 \text{ and } b(t) \neq 0.$$

Since x and t are independent variables, we must have

$$a''(x)/a(x) = b'(t)/b(t) = \lambda \text{ (some constant).}$$

This is possible if $a(x)$ and $b(t)$ satisfy

$$a''(x) - \lambda a(x) = 0 \quad \text{and}$$

$$b'(t) - \lambda b(t) = 0\,;$$

with boundary conditions

$$a(0)b(t) = 0\,,\ t > 0, \quad \text{and}$$

$$a(1)b(t) = 0\,,\ t > 0,$$

we will also have $u(0,t) = u(1,t) = 0,\ t > 0$.

Of course, $a(x) = 0,\ 0 < x < 1$, satisfies the differential equation for $a(x)$; however, unless $f(x) = 0$ for all x, this does not help us solve the original problem.

For certain values of λ, the problem

$$a''(x) - \lambda a(x) = 0$$

$$a(0) = a(1) = 0$$

has non-trivial (not identically zero) solutions, namely when

$$\lambda = -(k\pi)^2\,,\ k = 1, 2, 3, \ldots\,.$$

In this case, $a(x) = \sin(k\pi x)$ satisfies the differential equation and the boundary conditions.

A corresponding solution for $b(t)$ satisfying

$$b'(t) + (k\pi)^2 b(t) = 0$$

is
$$b(t) = e^{-(k\pi)^2 t}\, b(0)\,.$$

We find, thus, that

$$u(x,t) = b(0)e^{-(k\pi)^2 t} \sin(k\pi x)\,,\ k = 1, 2, 3, \ldots,$$

satisfies
$$u_t = u_{xx}\,,\ u(0,t) = u(1,t) = 0$$

for arbitrary $b(0)$.

Note that we can look upon the functions $\sin(k\pi x)$ as eigenfunctions of the linear operator $D^2 x = x''$ acting on twice differentiable functions which vanish at 0 and at 1, corresponding to the eigenvalues $\lambda = -(k\pi)^2, k = 1, 2, \ldots$.

Any linear combination of functions of the form above, say

$$u(x,t) = \sum_{k=1}^{\infty} b_k\, e^{-(k\pi)^2 t} \sin(k\pi x)$$

is also a formal solution of

$$u_t = u_{xx}, \ u(0,t) = u(1,t) = 0.$$

There is one further condition to be satisfied for a solution to the original problem, namely the initial condition $u(x, 0) = f(x), 0 \leqslant x \leqslant 1$. This is possible for the formal solution given above, if there exist constants $b_1, b_2, \ldots$ such that

$$\sum_{k=1}^{\infty} b_k \sin(k\pi x) = f(x), \ 0 \leqslant x \leqslant 1.$$

Suppose now that $g(0) = \bar{u}_0$ and $g(L) = \bar{u}_L$, then $f(0) = f(1) = 0$. In this case, we can find an extension of f to the interval $[-1, 1]$ such that $f(-x) = -f(x)$ for $x \in [0, 1]$.

We know that the functions 1, $\sin(k\pi x)$, $\cos(k\pi x)$, $k = 1, 2, \ldots$, form an orthonormal basis for $\mathcal{L}_2[-1, 1]$. Furthermore, we have $(f, 1) = 0$ and $(f, \cos(k\pi x)) = 0$ for this odd extension of f. Therefore,

$$f(x) = \sum_{k=1}^{\infty} (f, \sin(k\pi x)) \sin(k\pi x)$$

for every $f \in \mathcal{L}_2[-1, 1]$ with $f(0) = f(1) = 0$ and with

$$(u, v) = \int_{-1}^{1} u(t)v(t)\mathrm{d}t.$$

It is not hard to show that the functions $\sqrt{2}\sin(k\pi x)$, $k = 1, 2, \ldots$, form an orthonormal basis for $\mathcal{L}_2[0, 1] \cap \{f : f(0) = f(1) = 0\}$. Thus, if $g(\bar{x})$ in the original formulation of the problem is continuous with $g(0) = \bar{u}_0$ and $g(L) = \bar{u}_L$, then

$$\bar{u}(\bar{x}, \bar{t}) = \bar{u}_0 + (\bar{u}_L - \bar{u}_0)(\bar{x}/L) + u(x, t)$$

is the unique solution to the problem with

$$x = \bar{x}/L \text{ and } t = \overline{kt}/L^2 \text{ and with } b_k = 2\int_0^1 f(x)\sin(k\pi x)\mathrm{d}x.$$

At $t = 0$, the series converges pointwise to a continuous function $f(x)$ for every x in $[0, 1]$. If we truncate the series for $u(x, t)$ at $k = N$, we obtain the approximation

$$\bar{u}_N(\bar{x}, \bar{t}) = \bar{u}_0 + (\bar{u}_L - \bar{u}_0)(\bar{x}/L) + \sum_{k=1}^{N} b_k \exp(-(k\pi)^2\, \overline{kt}/L^2)\sin(k\pi\bar{x}/L).$$

For any $\bar{t} \geqslant 0$, $\bar{u}_N(\bar{x}, \bar{t})$ converges to the solution $\bar{u}(\bar{x}, \bar{t})$ for all $0 \leqslant \bar{x} \leqslant L$.

Exercise 81 For a fixed $\bar{t}$, we can view $\bar{u}_N$ and $\bar{u}$ as elements of $\mathcal{L}_2\,[0, L]$. Find an upper bound on $\|u - u_N\|$ as a function of t.

Exercise 82 Show that the two-point boundary value problem

$$a''(x) - \lambda\, a(x) = 0$$

$$a(0) = a(1) = 0$$

can be written as an integral equation of the form

$$a(x) = \lambda \int_0^1 K(x, t) a(t) \mathrm{d}t$$

with a Hilbert–Schmidt kernel $K(x, t)$. Show that the integral operator has eigenvalues $-1/(k\pi)^2$, $k = 1, 2, \ldots$ and corresponding eigenfunctions $\sin(k\pi x)$; that is, we can write the integral equation as $[A - (1/\lambda)I]\; a = 0$,

where $$(A\, a)\,(x) = \int_0^1 K(x, t) a(t) \mathrm{d}t\,.$$

Thus, $A : \mathcal{L}_2\,[0, 1] \rightarrow \mathcal{L}_2\,[0, 1]$ is self-adjoint and c. c. and does not have $+1$ as an eigenvalue. Show that A has an inverse defined on a dense subset of $\mathcal{L}_2\,[0, 1]$.

For an application of separation of variables and eigenfunction expansions to a partial differential equation of elliptic type with a moving, nonlinear boundary condition, see: R. E. Moore and L. M. Perko, 'Inviscid fluid flow in an accelerating cylindrical container', *J. Fluid Mechanics* **22**, part 2, 1965, pp. 305–320. See also Chapter 20.

Concerning *rates of convergence* of series expansions for orthonormal bases in separable Hilbert spaces, we have the following.

Let $\{e_n\}$ be an orthonormal basis in $\mathcal{L}_2[a, b]$. For any $f \in \mathcal{L}_2[a, b]$, we have

$$f = \sum_{n=1}^{\infty} a_n\, e_n \text{ with } a_n = (f, e_n) = \int_a^b f(t) e_n(t) \mathrm{d}t\,,$$

meaning that

$$\lim_{N\to\infty} \left\| f - \sum_{n=1}^{N} a_n\, e_n \right\| = 0, \ (\|x\| = (x, x)^{1/2})\,.$$

First of all, for *any* f in $\mathcal{L}_2[a, b]$, we have

$$\|f\|^2 = \sum_{n=1}^{\infty} a_n^2 \text{ and}$$

$$\left\| f - \sum_{n=1}^{N} a_n\, e_n \right\|^2 = \left(f - \sum_{n=1}^{N} a_n\, e_n,\, f - \sum_{n=1}^{N} a_n\, e_n \right)$$

$$= \|f\|^2 - 2 \sum_{n=1}^{N} a_n(f, e_n) + \sum_{n=1}^{N} a_n^2$$

$$= \|f\|^2 - \sum_{n=1}^{N} a_n^2$$

$$= \sum_{n=N+1}^{\infty} a_n^2 .$$

Thus, $$\left\| f - \sum_{n=1}^{N} a_n e_n \right\| = \left(\sum_{n=N+1}^{\infty} a_n^2 \right)^{1/2} \to 0 .$$

The *rate* at which a_n tends toward zero will determine the rate at which the partial sums converge toward f in the Hilbert space norm. We can get an indication of the dependence of these rates on the smoothness of the function f as follows.

For the orthonormal basis $\{e_n\}$ with

$$e_n(t) = e^{i2\pi nt}, \; i = \sqrt{-1}, \; (e_n, e_k) = \int_0^1 e_n(t) \overline{e_k(t)} \, dt,$$

in $\mathcal{L}_2[0, 1]$ over the *complex* scalar field, we have

$$f(t) = \sum_{n=-\infty}^{\infty} a_n e^{i2\pi nt}, \; a_n = \int_0^1 f(t) e^{-i2\pi nt} \, dt ,$$

and, if f has k continuous derivatives, then $f^{(k)}$ is in $\mathcal{L}_2[0, 1]$ and, by k-fold formal differentiation of the series for $f(t)$, we obtain

$$f^{(k)}(t) = \sum_{n=-\infty}^{\infty} (i2\pi n)^k a_n e^{i2\pi nt} .$$

From $(i2\pi n)^k a_n \to 0$, we have $a_n = o(n^{-k})$.

If the function f satisfies the periodicity properties

$$f^{(j)}(0) = f^{(j)}(1), \; j = 1, 2, 3, \ldots, k-1,$$

then, by k-fold integration by parts, we obtain

$$(i2\pi n)^k a_n = \int_0^1 f^{(k)}(t) e^{-i2\pi nt} dt = (i2\pi n)^k \int_0^1 f(t) e^{-i2\pi nt} dt.$$

In this case, the formal differentiation process produces the correct coefficients for $f^{(k)}(t)$.

For other (non-periodic) functions and other orthonormal bases, we may have different rates of convergence.

Exercise 83 Show that the two-point boundary value problem

$$x''(t) + a(t)x(t) = b(t)$$
$$x(0) = x(1) = 0$$

has a unique solution in $\mathcal{L}_2[0, 1]$ for any b in $\mathcal{L}_2[0, 1]$ if a is continuous in $[0, 1]$ and $\max_{t \in [0,1]} |a(t)| < \pi^2$. See Exercise 82.

Next we will discuss and compare some methods for solving linear operator equations of the form $(I - A)x = y$ where $A : X \to X$ is a bounded linear operator on an appropriate normed linear space X. Although it is not the most general condition under which the methods can be applied, we will make the additional assumption $\|A\| < 1$ in order to simplify this introductory discussion of a number of comparable methods.

We have already seen (Exercises 71, 82, 83) that we can put the two-point boundary value problem

$$x''(t) + a(t)x(t) = b(t)$$
$$x(0) = x(1) = 0$$

into the form $(I - A)x = y$, with an integral operator A such that $\|A\| < 1$ provided that $|a(t)|$ is not too large. We can also put the initial value problem

$$x'(t) = M(t)x(t) + b(t), \quad 0 \leqslant t \leqslant T,$$
$$x(0) \in E^n,$$

with $M(t)$ an n-by-n matrix and $b(t), x(t) \in E^n$, for T not too large, into the form $(I - A)x = y$ for $\|A\| < 1$; in fact,

$$x(t) = x(0) + \int_0^t b(s)\mathrm{d}s + \int_0^t M(s)x(s)\mathrm{d}s$$

by integrating both sides of the differential equation and incorporating the initial condition. We can put this into the form $(I - A)x = y$ by defining

$$y(t) = x(0) + \int_0^t b(s)\mathrm{d}s$$

$$(Ax)(t) = \int_0^t M(s)x(s)\mathrm{d}s.$$

If T is small enough so that

$$T \sup_{t \in [0, T]} \|M(t)\| < 1$$

then $\|A\| < 1$, if we operate in the linear space of continuous mappings of $[0, T]$ into E^n with a suitable norm.

The operators A, with $\|A\| < 1$, arising in the above ways are completely continuous.

We will discuss, in the remainder of this chapter, two types of *finite basis methods, Galerkin-type methods* and *collocation methods*, as well as *finite difference methods* for the approximate solution of problems of the two classes mentioned above (two-point boundary value problems and initial value problems for ordinary differential equations). Viewed in their integral equation forms, the problems are called *Fredholm integral equations of the second kind* or *Volterra integral equations* (for the integral equation formulation of the initial value problem).

I) Galerkin's method in Hilbert spaces

Consider $A : H \to H$ with A as an operator of one of the types described above, where H is a separable Hilbert space of suitable functions. Choose $\{e_i\}$ to be an orthonormal basis for H. Put

$$x^{(n)} = \sum_{i=1}^{n} c_i^{(n)} e_i$$

and find $c_i^{(n)}$ for $i = 1, 2, \ldots, n$ by solving the finite linear algebraic system

$$c_k^{(n)} - \sum_{i=1}^{n} c_i^{(n)} (Ae_i, e_k) = (y, e_k), k = 1, 2, \ldots, n .$$

Further details of the method depend on the choice of the $\{e_i\}$.

Exercise 84 Let H be the R.K.H.S., $H^{(1)}$ described in Chapter 9. Assume that $T = 1$ satisfies the required condition for the initial value problem. Let $0 \leqslant t_1 < t_2 < \ldots < t_n \leqslant 1$. Discuss the Galerkin method for $e_i = R_{t_i}$, $i = 1, 2, \ldots, n$ and also for $e_i = AR_{t_i}$, $i = 1, 2, \ldots, n$ for each of the two classes of problems (initial and boundary value problems).

II) Collocation methods

Consider $A : X \to X$ with A as before and where X is a suitable normed linear space. Choose the finite elements $e_1, e_2, \ldots, e_n$ to be linearly independent elements of X. Put

$$x^{(n)} = \sum_{i=1}^{n} c_i^{(n)} e_i$$

with $c_i^{(n)}, i = 1, 2, \ldots, n$, to be determined from

$$[(I - A)x^{(n)}] (t_k) = y(t_k), \; k = 1, 2, \ldots, n$$

for some $0 \leqslant t_1 < t_2 < \ldots < t_n \leqslant 1$.

Thus, the coefficients are to be found by solving the finite linear algebraic system

$$\sum_{i=1}^{n} [e_i(t_k) - (Ae_i)(t_k)]c_i^{(n)} = y(t_k),\ k = 1, 2, \ldots, n.$$

Exercise 85 Show that if $X = H^{(1)}$ as in Exercise 84, then with $e_i = R_{t_i}$, the Galerkin method is also a collocation method.

For a discussion of such methods in the R.K.H.S. described at the end of Chapter 9, with high rates of convergence, see: G. Wahba, 'A class of approximate solutions to linear operator equations' *J. Approx. Theory* **9**, 61–77, 1973.

Exercise 86 Let $X = C[0, 1]$; suppose that $y'' \in C[0, 1]$ and let e_i, $i = 1, 2, \ldots, n$ be the so-called 'hat functions'†

$$e_i(t) = \begin{cases} 0 & \text{for } t \leqslant t_{i-1} \text{ or } t \geqslant t_{i+1}. \\ (t - t_{i-1})/(t_i - t_{i-1}) & \text{for } t_{i-1} \leqslant t \leqslant t_i, \\ (t_{i+1} - t)/(t_{i+1} - t_i) & \text{for } t_i \leqslant t \leqslant t_{i+1}, \end{cases}$$

with $t_0 = 0 < t_1 < t_2 < \ldots < t_n < 1 = t_{n+1}$.

Let $\{c_i^{(n)}\}$ be determined by the finite linear algebraic system for the collocation method. Find an error bound (upper bound) on

$$\|x - x^{(n)}\| = \sup_{t \in [0,1]} |x(t) - x^{(n)}(t)|.$$

III) Finite difference methods

We can find approximate solution values $\{x(t_i)\}$ on a finite set of arguments $\{t_1, t_2, \ldots, t_n\}$ for problems in differential or integral equations by replacing derivatives by difference quotients and integrals by finite sums. For example, we can approximate $x'(t_i)$ by

$$[x(t_i + h) - x(t_i)]/h, \text{ 'forward difference scheme'},$$

or by $$[x(t_i + h) - x(t_i - h)]/2h), \text{ 'centred difference scheme'}.$$

We can approximate $x''(t_i)$ by

$[x(t_i + h) - 2x(t_i) + x(t_i - h)]/h^2$; and we can approximate

$$\int_a^b f(t)\,dt \text{ by } \sum_{j=1}^{n} w_j f(t_j)$$

† This is an example of a so-called 'finite element method', in which the basis functions vanish outside some finite (bounded) domain.

for some set of 'weights' $w_1, w_2, \ldots, w_n$ and arguments $t_1, \ldots, t_n$. We can also approximate partial derivatives by difference quotients; thus, for example, we can approximate the Laplacian, $u_{xx} + u_{yy}$, at (x_i, y_j) by

$$[u(x_i + h, y_j) - 2u(x_i y_j) + u(x_i - h, y_j)]/h^2 + [u(x_i, y_j + k) - 2u(x_i, y_j) + u(x_i, y_j - k)]/k^2 .$$

For an excellent treatment of finite difference methods in differential equations which combines techniques of classical analysis and techniques of functional analysis, see: I. Babuska, M. Prager, and E. Vitasek, *Numerical processes in differential equations*, Interscience, N.Y., 1966. Finite difference methods for integral equations are usually called 'quadrature methods', see: P. M. Anselone, *Collectively compact operator approximation theory and applications to integral equations*, Prentice-Hall, N. J., 1971.

For a linear initial value problem or a linear boundary value problem in differential equations or for a linear integral equation, by replacing derivatives or integrals by difference quotients or by finite sums, we will obtain a finite dimensional system of linear algebraic equations for the approximate solution values at a finite set of argument values. If the matrix of coefficients of the resulting system is non-singular, we can find (or approximate) the unique solution for the approximate solution values by Gaussian elimination or other means.

We can find error bounds on the discrete solutions obtained from finite difference methods in a variety of ways. One approach is to analyze the effect of the approximation of the derivatives by finite difference quotients (and integrals by finite sums) upon errors in the solution by using comparisons of finite Taylor series expansions with remainder terms in mean value form or in integral form. Another approach is to interpolate the discrete solution with a piecewise linear continuous function (or a smoother interpolating function such as a cubic spline function, etc.). We can then take the interpolating function to be the x_0 in an iterative method for solving the problem in integral equation form and obtain error bounds by the method for analyzing the convergence of the partial sums in a Neumann series.

To illustrate, suppose that $A : C[0, 1] \to C[0, 1]$ with $\|A\| < 1$. Let $x_0(t)$ interpolate the values of a discrete solution of a finite difference equation. Suppose that we can put the original problem in the form $(I - A)x = y$ for y in $C[0, 1]$. Then we can consider the iterative scheme $x_{p+1} = y + Ax_p, p = 0, 1, 2, \ldots$ and we can obtain bounds on the errors in $x_0(t_i)$ as follows.

Suppose that the finite difference method produces numerical values $\bar{x}_i$, $i = 1, 2, \ldots, n$, approximating the exact values of the solution $x(t_i)$. The exact solution satisfies $x = y + Ax$. We suppose that $x_0(t_i) = \bar{x}_i$ for all i, and that $x_0 \in C[0, 1]$.

The norm in $C[0, 1]$ leads to the result

$$|\bar{x}_i - x(t_i)| \leqslant \|x_0 - x\| \quad \text{for all } i = 1, 2, \ldots, n.$$

In order to bound the right hand side of the above inequality, we can proceed as follows. From

$$x = y + Ax\ , \text{ and}$$

$$x_1 = y + Ax_0\ ,$$

we obtain $\|x - x_1\| \leqslant \|A\|\ \|x - x_0\|$.

Now $\quad x - x_0 = x - x_1 + x_1 - x_0$,

so $\quad \|x - x_0\| \leqslant \|x - x_1\| + \|x_1 - x_0\|$.

Combining the two inequalities above, we obtain

$$\|x - x_0\| \leqslant (\|x_1 - x_0\|)/(1 - \|A\|)\ .$$

Furthermore, if we can compute x_1 and bound $\|x_1 - x_0\|$, we will have the sharper bound (since $\|A\| < 1$) on the error in x_1:

$$\|x - x_1\| \leqslant [\|A\|\ \|x_1 - x_0\|]\ /(1 - \|A\|)\ .$$

For a specific example of the above procedure, consider the following initial value problem:

$$x'(t) = tx(t) + 2$$

$$x(0) = 1.$$

We can apply a finite difference method to obtain an approximate value of the solution at $t = 0.1$. Suppose that we obtain $\bar{x} = 1.2057$ by some finite difference method as our approximation to the exact solution at $t = 0.1$. We now wish to bound the error in $\bar{x}$.

From the given differential equation, we find that the exact solution must have an initial slope of $x'(0) = 2$. Thus, we can look at quadratic polynomials of the form $p(t) = 1 + 2t + at^2$, all of which satisfy the given initial condition $x(0) = 1$ and agree with the initial slope of the solution. We can solve for the value of a for which $p(t)$ agrees with $\bar{x}$ when $t = 0.1$. We find the value $a = 0.57$. Thus, if we put $x_0(t) = 1 + 2t + 0.57t^2$, then $x_0(t)$ interpolates the finite difference solution at $t = 0.1$ and is an element of $C[0, 1]$. Actually, in this illustration, since we are only interested in what happens for t up to 0.1, it is better to regard x_0 and x as elements of $C[0, 0.1]$ with the norm

$$\|x\| = \max_{t \in [0, 0.1]} |x(t)|.$$

In this way, we can find sharper bounds on the error in an approximation to the solution at $t = 0.1$. We do not have to bound anything that goes beyond the point $t = 0.1$.

We can rewrite the initial value problem in the form $(I - A)x = y$, by integrating both sides of the differential equation and applying the initial condition; thus, we have

$$x(t) = 1 + 2t + \int_0^t sx(s)\,ds\ .$$

Now, we put $y(t) = 1 + 2t$ and define the linear operator A by

$$(Ax)(t) = \int_0^t sx(s)\,ds\ .$$

Thus, we have $x = y + Ax$ or $(I - A)x = y$ as desired.

For the operator $A : C[0, 0.1] \to C[0, 0.1]$ defined above, we find that

$$\|A\| = \sup_{\|x\|=1} \left(\max_{t\in[0,0.1]} \left| \int_0^t sx(s)\,ds \right| \right) \leqslant 0.005\ .$$

From $x_0(t) = 1 + 2t + 0.57t^2$, using $x_1 = y + Ax_0$, we find

$$x_1(t) = 1 + 2t + t^2/2 + 2t^3/3 + 0.57t^4/4.$$

It is not difficult to show that

$$\|x_1 - x_0\| = \max_{t\in[0,0.1]} |x_1(t) - x_0(t)| < 0.00012\ .$$

It follows that

$$\|x - x_0\| < 0.0001206$$

and $$\|x - x_1\| < 0.00000061\ .$$

In particular, we have found that, for $\bar{x} = 1.2057$, we have the error bound

$$|\bar{x} - x(0.1)| < 0.0001206\ ;$$

and for $x_1(0.1) = 1.2056809$, we have

$$|x_1(0.1) - x(0.1)| < 0.00000061\ .$$

Thus, the exact value of the solution at $t = 0.1$ satisfies

$$1.205680 < x(0.1) < 1.205682\ .$$

The best existing mathematical software for initial value problems in ordinary differential equations is discussed in: G. Corliss and Y.F. Chang, *Solving ordinary differential equations using Taylor series*, ACM Transactions on Mathematical Software, Vol. 8, No. 2, 1982, 114–144.

Exercise 87 Consider a linear Fredholm integral equation of the second kind

$$f(x) - \int_a^b K(x, y) f(y) \mathrm{d}y = g(x) .$$

Discuss methods of *Nystrom type* (also known as *quadrature methods*) for this equation:

1) choose a quadrature formula for approximating integrals, of the form

$$\int_a^b h(y) \mathrm{d}y = \sum_{j=1}^{n} w_j h(y_j) + E_n(h)$$

with $E_n(h) \to 0$ as $n \to \infty$;

2) obtain an approximate solution to the integral equation of the form

$$f_n(x) = g(x) + \sum_{j=1}^{n} w_j K(x, y_j) f_n(y_j)$$

by solving the linear system

$$f_n(y_i) = g(y_i) + \sum_{j=1}^{n} w_j K(y_i, y_j) f_n(y_j) , \quad i = 1, 2, \ldots, n,$$

for $f_n(y_i), i = 1, 2, \ldots, n$.

Obtain an upper bound on $\| f - f_n \|$ in an appropriate function space. Discuss conditions under which the method converges.

It should be pointed out that finite difference methods are not restricted to linear operator equations by any means. There is, in fact, an enormous literature on finite difference methods for both linear and nonlinear problems in ordinary and partial differential equations. Much of the analysis of such methods, however, is based on classical real analysis and not on functional analysis (although some of it is). In any case, we will leave this subject, important as it is to numerical solution of various problems in operator equations, to other works. It is, in the opinion of the author, outside the scope of this introductory volume on numerical functional analysis.

CHAPTER 14

Interval methods for operator equations

Recall, from Chapter 10, the definitions of interval valued mappings in partially ordered spaces of functions, set valued extensions, united extensions, etc.

We will illustrate, in this chapter, the application of the theorem in Chapter 10 to the construction of nested sequences of intervals of functions containing exact solutions to operator equations. The methods will apply to nonlinear operator equations as well as to linear operator equations.

In later chapters we will discuss other methods for nonlinear operator equations.

For an introduction to the methods and results of *interval analysis*, see: R. E. Moore, *Methods and applications of interval analysis*, SIAM Studies in Applied Mathematics, SIAM, Philadelphia, 1979.

For the methods to be discussed in this chapter, some additional background material is needed. Specifically, we will use *interval arithmetic* and *interval integration*.

Interval arithmetic operations are defined as the united extensions of real arithmetic operations to pairs of intervals. Thus,

$$\begin{aligned}
[a, b] + [c, d] &= \{x + y : x \in [a, b], y \in [c, d]\} \\
&= [a + c, b + d], \\
[a, b] - [c, d] &= \{x - y : x \in [a, b], y \in [c, d]\} \\
&= [a - d, b - c], \\
[a, b]\,[c, d] &= \{x y : x \in [a, b], y \in [c, d]\} \\
&= [\min(ac, ad, bc, bd), \max(ac, ad, bc, bd)],
\end{aligned}$$

and

$$\begin{aligned}
[a, b] / [c, d] &= \{x/y : x \in [a, b], y \in [c, d]\} \\
&= [a, b]\;[1/d, 1/c] \quad \text{for } 0 \notin [c, d].
\end{aligned}$$

With these definitions, we can carry out sequences of arithmetic operations on closed, bounded intervals on the real line as long as division by an interval containing zero does not occur. A more general definition of division by an interval exists and has important applications to root finding and to optimization (mathematical programming), but it will not be needed here. By identifying a real number with a degenerate interval $[a, a]$, we can see that these definitions of interval arithmetic operations are, indeed, extensions of ordinary real arithmetic operations. For a study of properties of this arithmetic system, see the reference given above in this chapter. It is important to remember that, for an interval $[a, b]$, we always have $a \leqslant b$.

If f and g are two real valued functions on a domain D, we write $f \leqslant g$ iff $f(t) \leqslant g(t)$ for all t in D. We denote by $[f, g]$ the interval of functions h such that $f(t) \leqslant h(t) \leqslant g(t)$ for all t in D.

Now let f and g be real valued functions, defined on an interval $[a, b]$ of real numbers, and such that $f \leqslant g$. Let F be the interval valued function defined by $F(t) = [f(t), g(t)]$. The definition of interval integration we will give can be extended to functions with values on the extended real line, but we will not use this here. See: O. Caprani, K. Madsen, and L. B. Rall, 'Integration of interval functions', *SIAM J. Math. Anal.* **12**, 321–341 (1981). We define the *interval integral* (integral of an interval valued function, in general interval valued itself) of F over $[a, b]$ by:

$$\int_a^b F(t)\mathrm{d}t = \left[\underline{\int_a^b} f(t)\mathrm{d}t,\ \overline{\int_a^b} g(t)\mathrm{d}t\right],$$

where the integrals on the right hand side of the definition are the lower Darboux integral of f and the upper Darboux integral of g. These integrals exist for *any* real valued functions f and g.

It can be shown (see reference in the above paragraph) that if F itself is real valued (degenerate interval valued), then the interval integral of F is a real number if F is Riemann integrable; otherwise the integral may be interval valued.

It is important to mention some properties of the interval integral. First, if $h \in F$ (that is, $h(t) \in F(t)$ for all t in $[a, b]$), then, for the real valued function h, we have

$$\int_a^b h(t)\mathrm{d}t \in \int_a^b F(t)\mathrm{d}t$$

where the integral on the left can be regarded as either a Riemann integral or a Lebesgue integral (if these exist), or even as an interval integral for $h(t) = [h(t), h(t)]$. Second, if F and G are two interval valued functions with $F \subseteq G$ (that is, $F(t) \subseteq G(t)$ for all t in $[a, b]$), then

$$\int_a^b F(t)\mathrm{d}t \subseteq \int_a^b G(t)\mathrm{d}t\,.$$

Thus, interval integration preserves inclusion.

Let s be a positive real number. Since we identify s with the degenerate interval $[s, s]$, it follows from the definition of an interval product that $s[a, b] = [sa, sb]$ for any interval $[a, b]$. Now let $A_0, A_1, \ldots, A_n$ be intervals of real numbers. Consider the polynomial $P(t) = A_0 + A_1 t + \ldots + A_n t^n$, where t is a real variable. We can integrate such a polynomial with the definition of an interval integral given above and obtain the explicit representation

$$\int_0^x P(t)\mathrm{d}t = A_0 x + A_1 x^2/2 + \ldots + A_n x^{n+1}/(n+1), \ x > 0.$$

Let us now reconsider the initial value problem discussed at the end of the previous chapter

$$x'(t) = tx(t) + 2$$
$$x(0) = 1.$$

We rewrite this, as before, as the integral equation

$$x(t) = 1 + 2t + \int_0^t sx(s)\mathrm{d}s .$$

It is clear from the initial value formulation of the problem that there is a unique solution for all real t (from the elementary theory of differential equations). In fact, the solution is

$$x(t) = \exp(t^2/2) \left(1 + 2 \int_0^t \exp(-s^2/2)\,\mathrm{d}s\right).$$

In any case, we can expect a power series for the solution (expanded about $t = 0$) to begin with the terms

$$x(t) = 1 + 2t + ct^2 + \ldots$$

for some c.

With this much inspection of the problem, we choose an interval polynomial of the form

$$X_0(t) = 1 + 2t + [a, b]t^2 ,$$

for some interval $[a, b]$ to be determined; we define the *interval operator* P by

$$P(X)(t) = 1 + 2t + \int_0^t sX(s)\mathrm{d}s \text{ ; and we compute}$$

$$P(X_0)(t) = 1 + 2t + \int_0^t s(1 + 2s + [a, b]s^2)\mathrm{d}s$$

$$= 1 + 2t + t^2/2 + (2/3)t^3 + [a/4, b/4]\,t^4 .$$

We will have $P(X_0)(t) \subseteq X_0(t)$ if

$$1 + 2t + t^2/2 + (2/3)t^3 + [a/4, b/4]t^4 \subseteq 1 + 2t + [a, b]t^2.$$

The inclusion above will hold if

$$1/2 + (2/3)t + [a/4, b/4]t^2 \subseteq [a, b].$$

It follows that we will have

$$P(X_0)(t) \subseteq X_0(t) \text{ for all } t \text{ in } [0, 0.1]$$

provided that

$$a \leqslant \tfrac{1}{2} \text{ and } 0.56808688\ldots \leqslant b.$$

We can take $a = \frac{1}{2}$ and $b = 0.56808689$; then we find that $P(X_0)(t) \subseteq X_0(t)$ for all t in $[0, 0.1]$. It follows from the theorem in Chapter 10 that if the solution is in $X_0(t)$ it is also in $P(X_0)(t)$ for t in $[0, 0.1]$. In fact, it is not hard to show that, for this example, the solution is in $X_0(t)$ for the chosen values of a and b for all t in $[0, 0.1]$; see section 8.1 of the first reference in this chapter. It follows that the solution satisfies

$$x(t) \in P(X_0)(t) \subseteq 1 + 2t + t^2/2 + (2/3)t^3 + [0.125, 0.142022]t^4$$

for all t in $[0, 0.1]$.

In particular, we find from this method that

$$1.205679 \leqslant x(0.1) \leqslant 1.205681.$$

Compare this result with the one obtained at the end of the previous chapter on this same example. The two results, obtained by quite different methods, are comparable.

We could, in fact, *intersect* the two intervals and conclude that $1.205680 \leqslant x(0.1) \leqslant 1.205681$. Such an operation is quite typical of the approach of interval analysis.

We can continue the interval valued function we have found, containing the solution, beyond $t = 0.1$. From $x'(t) = tx(t) + 2$, we have

$$x(t) = x(0.1) + 2(t - 0.1) + \int_{0.1}^{t} sx(s)\,ds.$$

Now let $A_0 = [1.205679, 1.205681]$, the interval we have already found which contains $x(0.1)$. Let $X_0(t)$ be the interval polynomial

$$X_0(t) = A_0 + 2(t - 0.1) + [a, b](t - 0.1)^2,$$

with $[a, b]$ to be determined, if possible, so that

$$P(X_0)(t) \subseteq X_0(t) \quad \text{for all } t \text{ in, say, } [0.1, 0.2],$$

where we put

$$P(X)\,(t) = A_0 + 2(t-0.1) + \int_{0.1}^{t} sX(s)\,ds\,.$$

Evaluating the interval operator P for $X = X_0$, we obtain

$$P(X_0)\,(t) = A_0 + 2(t-0.1) + (A_0/2)\,(t-0.1)^2 + (2/3)\,(t-0.1)^3 + [a/4, b/4]\;(t-0.1)^4\,.$$

We find that $P(X_0)\;(t) \subseteq X_0(t)$ for all t in $[0.1,\ 0.2]$ provided that $a \leqslant 0.6028395$ and $0.67118513 \leqslant b$. It follows that the solution satisfies

$$\begin{aligned} x(t) \in {} & [1.205679, 1.205681] + 2(t-0.1) \\ & + [0.6028395, 0.6028405]\;(t-0.1)^2 \\ & + (2/3)\,(t-0.1)^3 + [0.15070987, 0.16779629]\;(t-0.1)^4 \end{aligned}$$

for all t in $[0.1, 0.2]$. In particular, we have

$$1.412389 \leqslant x(0.2) \leqslant 1.412393\,.$$

Exercise 88 Verify the above calculations and then use the same procedure to extend the interval bounds on the solution to the interval $[0.2, 0.3]$.

It is easy to show that interval arithmetic is *inclusion isotone*. From this and from the fact that interval integration preserves inclusion, it follows that the sequence of interval valued functions defined by $X_{k+1}(t) = P(X_k)\,(t)$, $k = 0, 1, 2, \ldots$ is a nested sequence provided that $P(X_0)\,(t) \subseteq X_0(t)$ for all t in some interval $[t_1,\ t_2]$, where P is either of the interval operators defined for the example above for the intervals $[0, 0.1]$ or $[0.1, 0.2]$. In fact, it can be shown that, for the two sequences of real valued functions $\{\underline{X}_k\}$ and $\{\overline{X}_k\}$ where $X_k(t) = [\underline{X}_k(t), \overline{X}_k(t)]$, we have the following:

1) the solution of the initial value problem satisfies

$$\underline{X}_k(t) < x(t) < \overline{X}_k(t) \text{ for all } t \text{ in } [t_1,\ t_2] \text{ and all } k;$$

2) the sequence $\{\underline{X}_k(t)\}$ converges pointwise to $x(t)$ and is monotonic increasing;
3) the sequence $\{\overline{X}_k(t)\}$ converges pointwise to $x(t)$ and is monotonic decreasing.

Similar results can be obtained for a very large class of initial and boundary value problems, both linear and nonlinear, as shown in the first reference in this chapter. We will illustrate the method further by giving additional examples.

In the preceding example, we made use of a property of interval arithmetic which was not explicitly stated, namely the cancellation property of interval addition: if $A + X = A + Y$ for intervals A, X, and Y, then $X = Y$. This property, which follows directly from the definition of interval addition, also implies that if $A + X \subseteq A + Y$, then $X \subseteq Y$. The latter implication has been used several times in the previous example.

Another comment which might be in order is this. To say that interval arithmetic is inclusion isotone means that, if $X \subseteq X'$ and $Y \subseteq Y'$, then

$$X + Y \subseteq X' + Y' ,$$
$$X - Y \subseteq X' - Y' ,$$
$$X\,Y \subseteq X'\,Y' , \text{ and}$$
$$X/Y \subseteq X'/Y' .$$

Note that if $0 \notin Y'$ and $Y \subseteq Y'$, then $0 \notin Y$. These properties also follow directly from the definitions of the interval arithmetic operations.

Consider next the following two-point boundary value problem.

$$x''(t) + \exp(-t^2)\,x(t) = 1$$
$$x(0) = x(1) = 0.$$

By integrating twice and imposing the boundary conditions, we can rewrite this problem as the following integral equation

$$x(t) = t(t-1)/2 + \int_0^t s(1-t)\exp(-s^2)\,x(s)\,\mathrm{d}s$$
$$+ \int_t^1 t(1-s)\exp(-s^2)\,x(s)\,\mathrm{d}s .$$

Using the Taylor series with remainder, we can write down interval polynomials which contain the exponential function as follows. For all s in $[0, 1]$ and for any integer $k \geqslant 0$,

$$\exp(-s^2) \in A_k(s) = 1 - s^2 + \ldots + (-s^2)^k/k!$$
$$+ \left\{(-s^2)^{k+1}/(k+1)!\right\}\,[\mathrm{e}^{-1}, 1].$$

Note that the last term has an interval coefficient. The result just given follows from

$$\mathrm{e}^{-u} = \sum_{i=0}^{k} (-u)^i/i! + \left\{(-u)^{k+1}/(k+1)!\right\}\mathrm{e}^{-v} \text{ for some } v \in [0, u]$$

by putting $u = s^2$ with $s \in [0, 1]$, and replacing v by $[0, 1]$.

We now define the interval operator P_k by

$$P_k(X)(t) = t(t-1)/2 + \int_0^t s(1-t)A_k(s)X(s)\,\mathrm{d}s$$

$$+ \int_t^1 t(1-s)A_k(s)X(s)\mathrm{d}s \,.$$

We seek an interval valued function X_0 such that

$$P_k(X_0)(t) \subseteq X_0(t) \text{ for all } t \text{ in } [0, 1].$$

From a result in a previous chapter, we know that a solution does exist since $\max_{t \in [0,1]} |e^{-t^2}| < 8$. (See Exercise 71.)

In the computations for this problem we will need the following additional properties of interval arithmetic; see the first reference in this chapter. For intervals A, B, and C, we have the *sub-distributive property* $A(B + C) \subseteq AB + AC$. Furthermore, for $a > 0$, we have $aA + aB = a(A + B)$; and, for $t \in [0, 1]$, we have $tC + (1 - t)C = C$.

Exercise 89 Show that we can take, for $X_0(t)$, the constant interval valued function $X_0(t) = [-4/27, 0]$ and satisfy the requirement that $P_k(X_0)(t) \subseteq X_0(t)$ for all t in $[0, 1]$ when $k = 0$.

For $k = 0$ we find that $P_0(X_0)(t) \subseteq [71/81, 96/81]\,t(t - 1)/2$; and, therefore, $P_0(X_0)(t)$ is contained in $X_0(t)$ if we take $X_0(t)$ to be the constant interval valued function $[-4/27, 0]$ for all t in $[0, 1]$. Even with this very crude approximation to the exponential function, we obtain the result that the exact solution value at $t = 0.5$ satisfies $0.109 < x(0.5) < 0.1482$. By iterating $X_{i+1}(t) = P_k(X_i)(t)$ and by using higher values of k for the series approximation $A_k(t)$ to the exponential, we can obtain arbitrarily sharp upper and lower bounds to the exact solution for all t in $[0, 1]$.

We next consider a nonlinear initial value problem. We add the further complications of *data perturbations* and a right hand side that does not satisfy the usual Lipschitz condition required for a unique solution. Nevertheless, the same method we have been discussing applies and will find an interval valued function containing *all* the solutions in a single interval computation.

We consider the initial value problem

$$y' = ct^2 + y^b, \quad y(0) = a,$$

where all we know about a, b and c is that

$$0 \leqslant a \leqslant 0.1, \quad 0.2 \leqslant b \leqslant 0.38, \quad \text{and} \quad 3.3 \leqslant c \leqslant 3.6.$$

Note that the right hand side of the differential equation does not satisfy a Lipschitz condition in y when $a = 0$.

We can write the initial value problem as an integral equation

$$y(t) = a + \int_0^t (cs^2 + y(s)^b)\mathrm{d}s \,.$$

Thus, we define the interval operator

$$P(Y)(t) = [0, 0.1] + \int_0^t \{[3.3, 3.6]s^2 + Y(s)^{[0.2, 0.38]}\} \, ds.$$

We let $Y_0(t) = [0, B]$ for all t in $[0, t_1]$ with $B > 0$ and $t_1 > 0$ to be determined, if possible, such that $P(Y_0)(t) \subseteq Y_0(t)$ for all t in $[0, t_1]$.

Now we define

$$[0, B]^{[0.2, 0.38]} = \{y^b : y \in [0, B], b \in [0.2, 0.38]\}$$
$$= \begin{cases} [0, B^{0.2}], & 0 < B \leqslant 1 \\ [0, B^{0.38}], & 1 \leqslant B. \end{cases}$$

Similarly, for the interval valued function $Y(s)^{[0.2, 0.38]}$ which occurs in the integral above, we put, for $Y(s) = [\underline{Y}(s), \overline{Y}(s)]$,

$$Y(s) = \{y^b : y \in Y(s), b \in [0.2, 0.38]\}$$
$$= \begin{cases} [\underline{Y}(s)^{0.38}, \overline{Y}(s)^{0.2}] & \text{if } 0 \leqslant \underline{Y}(s) \leqslant \overline{Y}(s) < 1, \\ [\underline{Y}(s)^{0.38}, \overline{Y}(s)^{0.38}] & \text{if } 0 \leqslant \underline{Y}(s) \leqslant 1 \leqslant \overline{Y}(s), \text{ and} \\ [\underline{Y}(s)^{0.2}, \overline{Y}(s)^{0.38}] & \text{if } 1 < \underline{Y}(s) \leqslant \overline{Y}(s). \end{cases}$$

We find that

$$P(Y_0)(t) = [0, 0.1] + [1.1, 1.2]t^3 + [0, B]^{[0.2, 0.38]}t,$$

for $Y_0(t) = [0, B]$.

If we take $B = 1$, we find that $P(Y_0)(t) \subseteq Y_0(t)$ for all t in $[0, t_1]$ if t_1 is any positive number such that $t_1 + 1.2t_1^3 \leqslant 0.9$.

The above inequality is satisfied, for instance, when $t_1 = 0.6$. We can conclude every solution of the initial value problem with a, b, and c in the given ranges of values satisfies

$$1.1t^3 \leqslant y(t) \leqslant 0.1 + t + 1.2t^3 \text{ for all } t \text{ in } [0, 0.6].$$

The bounds can be sharpened by iterating $Y_{k+1}(t) = P(Y_k)(t), k = 0, 1, 2, \ldots$. Furthermore, we can continue the bounding interval function beyond $t = 0.6$ by restarting the initial value problem at $t = 0.6$ with whatever interval value we find for $Y_k(0.6)$.

As another application of the interval method we have been discussing, consider the hyperbolic partial differential equation $u_{xy} = 1 + u(u_x + u_y)$. Suppose we seek a solution which satisfies the conditions $u(x, 0) = u(0, y) = 0$. We can rewrite the problem in the form of three integral equations as:

$$u_1(x, y) = \int_0^y (1 + u(x, r)(u_1(x, r) + u_2(x, r))) dr,$$

$$u_2(x,y) = \int_0^x (1 + u(z,y)\,(u_1(z,y) + u_2(z,y)))\,dz, \text{ and}$$

$$u(x,y) = \int_0^x \int_0^y (1 + u(z,r)\,(u_1(z,r) + u_2(x,r)))\,dr\,dz\ ,$$

where $u_1(x, y) = u_x(x, y)$ and $u_2(x, y) = u_y(x, y)$. Using interval methods, we can find upper and lower bounds on the solution in the positive quadrant $x, y \geqslant 0$, or in some part of it.

We can interpret the three integrals on the right hand sides of the integral equations above as interval integrals defining operators P_1, P_2, and P, mapping interval valued functions into interval valued functions. Motivated by the forms of these three operators, we put $U_1(x, y) = [a, b]\,y$, $U_2(x, y) = [a, b]\,x$, and $U(x, y) = [a, b]\,xy$, with the interval $[a, b]$ to be determined so that

$$P_1(U_1, U_2, U) \subseteq U_1\ ,$$

$$P_2(U_1, U_2, U) \subseteq U_1\ , \text{ and}$$

$$P(U_1, U_2, U) \subseteq U$$

in some part of the positive quadrant.

Substituting the chosen forms for U_1, U_2, and U into the integrals and carrying out the interval integrations, we obtain

$$P_1(U_1, U_2, U)(x,y) = y + [a, b]^2 (xy^3/3 + x^2y^2/2)\ ,$$

$$P_2(U_1, U_2, U)(x,y) = x + [a, b]^2\ (x^2y^2/2 + x^3y/3), \text{ and}$$

$$P(U_1, U_2, U)(x,y)\ = xy + [a, b]^2\ (x^2y^3/6 + x^3y^2/6)\ .$$

It is not hard to show that we can satisfy the required set inclusions in the square: x in $[0, 1/2]$ and y in $[0, 1/2]$, if we choose $a = 1$ and $b = 1.134$. It follows that the solution satisfies $u(x, y) \in xy + [0.1666, 0.2144]\ (x^2y^3 + x^3y^2)$ for all such x and y. In particular, $u(1/2, 1/2) \in [0.2604, 0.2634]$.

We can obtain arbitrarily sharp bounds by iterating the procedure. By replacing U_1, U_2, and U with the functions $P_1(U_1, U_2, U), P_2(U_1, U_2, U)$, and $P(U_1, U_2, U)$ just obtained, and with the values of a and b chosen above, we obtain, in one more iteration, the sharper bounds on the solution:

$$\begin{aligned} u(x,y) \in\ & xy + x^2y^2\ (x+y)/6 + x^3y^3\ ([1, 1.286](y^2/45 + xy/16 + x^2/45) \\ & + [0.1666, 0.2144](y^2/15 + [1, 1.286](xy^4/84 + x^2y^3/60) \\ & + xy/8 \\ & + [1, 1.286](x^2y^3/36 + x^3y^2/36) + x^2/15 \\ & + [1, 1.286](x^3y^2/60 + x^4y/84)))\ . \end{aligned}$$

At $x = y = 1/2$, the above expression, evaluated in interval arithmetic, produces the result

$$u(1/2, 1/2) \in [0.2610, 0.2612] \text{ , that is,}$$

$$0.2610 \leqslant u(1/2, 1/2) \leqslant 0.2612.$$

We could obtain arbitrarily sharp upper and lower bounds by iterating further and carrying enough digits.

Extensive software packages are available for carrying out interval computations on a wide variety of computers, for example, Pascal–SC (available from FBSoftware, 135 N. Prospect Ave., Madison, WI 53705, USA), and ACRITH (available from IBM as a set of FORTRAN subroutines). With these mathematical software packages, a user can write computer programs in Pascal or FORTRAN, treating intervals as a new data type. The details of interval arithmetic are carried out automatically by the software. Furthermore, the software includes problem-solving routines with guaranteed accuracy for a variety of problems such as: finding zeros of polynomials, solving linear algebraic systems, inverting matrices, linear optimization, eigenvalues and eigenvectors of matrices, and others. The ACRITH subroutine library, for instance, has been designed to operate under VM/SP in all IBM System/370 models 135 and above and the 303X, 308X, and 43XX processors. On the new 4361 processors, microcode support is provided in a high-accuracy arithmetic facility designed to make the interval arithmetic operations more efficient. Pascal-SC can be used on microcomputers such as Zilog, Apple, and IBM PC.

CHAPTER 15

Contraction mappings and iterative methods for operator equations in fixed point form

An operator equation is said to be *in fixed point form* (or it is called a *fixed point problem*) if it is written as $x = P(x)$, where $P: X \to X$ for some function space X. We have already seen operator equations of this type in Fredholm and Volterra integral equations, and we have converted initial and two-point boundary value problems in differential equations to integral equations of this type.

A solution x is called a *fixed point* of the operator P, since P maps the point x into itself.

Given an operator equation of the form

$$Q(x) = y\,, \quad x \in X, \quad y \in Y,$$

where $Q: X \to Y$ is an operator mapping elements of X into elements of Y, it is often possible to find a mapping P such that a fixed point of P is a solution of the given operator equation. Thus, for example, for the linear operator equation $(I - A)x = y$, we can write $x = y + Ax$, so that, for $P(x) = y + Ax$, we have the equivalent formulation $x = Px$.

As another example, we can often convert an operator equation of the form $Q(x) = y$ into one of fixed point form, even when Q is a nonlinear operator. Suppose we can find a nonsingular linear operator $A: Y \to X$, where X and Y are linear spaces; thus, $Az = 0$ implies that $z = 0$. Then a fixed point of P, with P defined by $P(x) = x + A\,(Q(x) - y)$, $Q: X \to Y$, $y \in Y$, $P: X \to X$, is a solution of $Q(x) = y$.

Most methods for the approximate solution of nonlinear operator equations (and some methods for linear operator equations) are of iterative type; and most methods of iterative type have the form $x_{k+1} = P(x_k)$, $k = 0, 1, 2, \ldots$ for some operator P.

Note that even if a problem is already given in fixed point form $x = T(x)$, there are many ways we could choose another operator P for an interative method. For instance, we can put $Q(x) = x - T(x)$ and $P(x) = x + A(x)Q(x)$. So long as $A(x)$ remains nonsingular, we can consider the iterative method of the type given above for the approximate solution of the fixed point problem $x = T(x)$.

There are several things to consider in the study of any iterative method. One is how to choose the initial guess x_0. Another is: under what conditions will the method converge in some sense or other (strong convergence, weak convergence, pointwise convergence, uniform convergence, order convergence, etc.). Still another is: when to stop iterating and how to estimate the error in the approximate solution x_k if we stop at the kth iteration.

We will discuss these questions concerning iterative methods in this chapter and in some of the remaining chapters.

For linear fixed point problems, sufficient conditions for convergence will usually be independent of the initial guess. For nonlinear fixed point problems, on the other hand, convergence will usually depend heavily on x_0 being close to a solution.

Definition
Let S be a closed subset of a complete metric space X. An operator $P: S \to S$ is called a (strong) *contraction mapping* of S if there is a constant $c < 1$ such that $d(P(x), P(y)) \leqslant cd(x, y)$ for all x, y in S. Note that a contraction mapping is necessarily continuous.

The following theorem is one of many possible versions of the *contraction mapping theorem.*

Theorem Let X be a complete metric space and let S_0 be a closed subset of X of finite diameter

$$d_0 = \sup_{x,y \in S_0} d(x, y) < \infty .$$

Let $P: S_0 \to S_0$ be a contraction mapping. Then the sequence of iterates $\{x_k\}$ produced by successive substitution $x_{k+1} = P(x_k)$, $k = 0, 1, 2, \ldots$ converges to $x = P(x)$, the unique fixed point of P in S_0, for any x_0 in S_0. Furthermore $d(x, x_k) \leqslant c^k d_0$, if $d(P(x), P(y)) \leqslant cd(x, y)$ for all x, y in S_0.

Proof Define the sets S_k recursively by $S_{k+1} = P(S_k) = \{P(x): x \in S_k\}$; thus $x_0 \in S_0$ implies $x_k \in S_k$ for all k. By the subset property for an arbitrary mapping (See Chapter 10), it follows that $\{S_k\}$ is a nested sequence of subsets of S_0. Now define d_k to be the diameter of S_k,

$$d_k = \sup_{x,y \in S_k} d(x, y) .$$

Since P is a contraction mapping, we have $d_{k+1} \leqslant cd_k$, $k = 0, 1, 2, \ldots$. Thus, $d_k \leqslant c^k d_0$, $k = 0, 1, 2, \ldots$; and, since $c < 1$, we have $d_k \to 0$ as $k \to \infty$.

It is easy to show that, for any x_0 in S_0, the sequence of points $\{x_k\}$ is a Cauchy sequence and so converges to a point x in the complete metric space X. Since $d_k \to 0$, the limit point is unique and independent of the starting point x_0. Since S_0 is closed and $S_k \subseteq S_0$ for all k, the limit point x is in S_0. Since P is continuous, x is a fixed point of P. Since $x, x_k \in S_k$ for all k, we have finally, $d(x, x_k) \leqslant d_k \leqslant c^k d_0$ for all k; and the theorem is proved.

Exercise 90 Complete the above proof by showing that $\{x_k\}$ is a Cauchy sequence.

Exercise 91 Show that an alternative bound on the error in the kth iterate x_k which does not depend on d_0 is $d(x, x_k) \leqslant c^k d(x_0, x_1)/(1-c)$.

Exercise 92 Recall the sequence $\{x_n\}$ defined iteratively by $x_{n+1} = y - Ax_n$ in Exercise 70 in Chapter 11. Show that, for $\|A\| < 1$, the operator $P(x) = y - Ax$ is a contraction on the whole Banach space X. Apply the theorem just proved to Exercise 70.

We now illustrate the application of the contraction mapping theorem to some examples of operator equations.

Example 1
The real line R is a complete metric space with $d(x, y) = |x - y|$. The mapping $P: R \to R$ defined by $P(x) = 1 + x/3$ is a contraction on R, since $d(P(x), P(y)) = |x - y|/3$. Therefore, P has a fixed point in R. Now $S = [0, 1]$ is a closed subset of R and $d(P(x), P(y)) = |x - y|/3$ for all x and y in S; however, P does not map S into itself. In fact, $P(S) = [1, 4/3]$. Indeed, P does not have a fixed point in S.

Exercise 93 For what intervals $[a, b]$ does $P([a, b]) \subseteq [a, b]$ hold? What is the fixed point of P in Example 1 above?

Example 2
We can rewrite the initial value problem $x'(t) = f(t, x(t)), x(t_0) = x_0$ in the fixed point form

$$x(t) = P(x)(t) = x_0 + \int_{t_0}^{t} f(s, x(s))\,ds .$$

We will suppose that f is continuous in a suitable region. We can regard P as a mapping of the Banach space $C[t_0, t_1]$ into itself, with t_1 to be determined. Note that P is continuous if f is. Suppose now that we can find a positive real valued function M such that

$$\max_{\substack{t \in [t_0, t_1] \\ u \in [x_0 - r, x_0 + r]}} |f(t, u)| \leqslant M(r) .$$

for all $0 \leqslant r < r_1$, with r_1 to be determined. We define the set

$$S_r(x_0) = \{x : x \in C[t_0, t_1], \|x - x_0\| \leqslant r\} ,$$

where x_0 within the norm expression denotes the constant function $x(t) = x_0$ for all t in $[t_0, t_1]$. Note that $S_r(x_0)$ is a closed subset of $C[t_0, t_1]$ with finite diameter $2r$. It is easy to see that P will map $S_r(x_0)$ into itself provided that $(t_1 - t_0) M(r) \leqslant r$.

We next seek conditions under which P will be a contraction on $S_r(x_0)$. First, we find that, for x and y in $C[t_0, t_1]$, we have

$$\|P(x) - P(y)\| \leqslant (t_1 - t_0) \max_{t \in [t_0, t_1]} |f(t, x(t)) - f(t, y(t))| .$$

Next, we require that f satisfies a Lipschitz condition in $S_r(x_0)$: we suppose we can find $b(r) > 0$ such that, for all $|u - x_0| \leqslant r$ and $|v - x_0| \leqslant r$, we have $|f(t, u) - f(t, v)| \leqslant b(r) \, |u - v|$, $\forall \, t \in [t_0, t_1]$. With this condition, we then have $\|P(x) - P(y)\| \leqslant (t_1 - t_0) \, b(r) \, \|x - y\|$; and P will then be a contraction on $S_r(x_0)$ provided that $c = (t_1 - t_0) b(r) < 1$.

From this analysis, we can see that P will be a contraction mapping of $S_r(x_0)$ into itself for a given positive value of r if $(t_1 - t_0)$ is sufficiently small so that the two conditions

$$(t_1 - t_0) \, M(r) \leqslant r \quad \text{and} \quad (t_1 - t_0) \, b(r) < 1$$

are satisfied. In that case, we can apply the contraction mapping theorem to the *initial value problem* expressed in the fixed point form above.

To illustrate the details of such an application, we now consider a specific example of the form discussed.

For the initial value problem $x'(t) = [x(t)]^2$ with $x(0) = 1$, we have $f(t, u) = u^2$, in the notation of the general analysis for initial value problems given above. Thus, we seek $M(r)$ such that $|u^2| \leqslant M(r)$ whenever $u \in [1 - r, 1 + r]$. Thus, we can take $M(r) = (1 + r)^2$. For $b(r)$ such that $|f(t, u) - f(t, v)| = |u^2 - v^2| \leqslant b(r) \, |u - v|$, we can take $b(r) = 2(1 + r)$ for all $u, v \in [1 - r, 1 + r]$. Thus, P will be a contraction on $S_r(1)$ provided that $t_1 > 0$, $t_1 (1 + r)^2 \leqslant r$, and $c = 2t_1(1 + r) < 1$.

For any $r > 0$, these inequalities are satisfied for t_1 such that $0 < t_1 < \min \{r/(1 + r)^2, 1/(2(1 + r))\}$. It is not hard to show that $\frac{1}{4}$ is the least upper bound on admissible values of t_1.

Thus, for $r = 0.5$ and $t_1 = 0.2$, the mapping

$$P(x)(t) = 1 + \int_0^t [x(s)]^2 \, ds$$

is a contraction of $S_{0.5}(1)$ into itself.

We can take, as a starting point, the constant function $x_0(t) = 1$, which is in $S_{0.5}(1)$, and the iterative method

$$x_{k+1}(t) = P(x_k)(t) = 1 + \int_0^t x_k(s)^2 \, ds, \; k = 0, 1, 2, \ldots$$

will converge in the $C[0, 0.2]$ norm to the solution for $t \in [0, 0.2]$.

We find that

$$x_1(t) = 1 + \int_0^t 1 \, ds = 1 + t \; ;$$

thus, $$\|x_1 - x_0\| = \max_{t \in [0, 0.2]} |x_1(t) - x_0(t)| = 0.2 \; .$$

Since $c = 2t_1(1 + r) = 0.6$, we have, applying the general error bound given in Exercise 91, for all t in $[0, 0.2]$,

$$|x_k(t) - x(t)| \leqslant \|x_k - x\| \leqslant (0.6)^k \, (0.2/0.4) = (1/2) \, (0.6)^k \; .$$

Exercise 94 Define the interval mapping P by

$$P(X)(t) = 1 + \int_0^t X(s)^2 \, ds \text{ and put } X_0(t) = [a, b] \, .$$

Show that, for $a = 1$ and $b = 1.5$, we have $P(X_0)(t) \subseteq X_0(t)$ for all t in $[0, 0.2]$. It follows from the results of Chapters 10 and 14 that the interval functions defined iteratively by $X_{k+1}(t) = P(X_k)(t)$ contain the solution for all k and all t in $[0, 0.2]$. Find $X_1(t)$ and $X_2(t)$ explicitly and compare these results with those obtained from the contraction mapping theorem.

Example 3

Consider the nonlinear *two-point boundary value problem* $x''(t) = e^{-x(t)}$, $x(0) = x(1) = 0$. We can put this problem in fixed point form as

$$x(t) = P(x)(t) = \int_0^1 K(t, s) e^{-x(s)} \, ds$$

with $$K(t, s) = \begin{cases} (t-1)s \, , & 0 \leqslant s \leqslant t \leqslant 1, \\ (s-1)t \, , & 0 \leqslant t \leqslant s \leqslant 1. \end{cases}$$

We can view P as a mapping $P: C[0, 1] \to C[0, 1]$. We define

$$S_r = \{ x : x \in C[0, 1], -r \leqslant x(t) \leqslant 0, \forall \, t \in [0, 1] \} \, .$$

For any positive r, S_r is a closed subset of $C[0, 1]$ with finite diameter r.

It is not hard to show that P maps S_r into itself if $-r \leqslant -e^r/8$.

The above inequality is true, for instance, if $r = 0.15$. Furthermore, for all x, y in S_r, we have $\|P(x) - P(y)\| \leqslant (e^r/8)\,\|x - y\|$. Thus, P is a contraction on S_r if $r \geqslant e^r/8$ and $e^r/8 \leqslant c < 1$; this is true for $r = 0.15$, and then we can take, for the value of c in the contraction mapping theorem, $c = 0.14523$.

There is one solution in $S_{0.15}$, and the iterative method $x_{k+1} = P(x_k)$ converges to it from any x_0 in $S_{0.15}$; for instance for $x_0(t) = 0$ for all t in $[0, 1]$.

It can be shown that there is a second solution to this nonlinear boundary value problem, which is not in $S_{0.15}$.

Exercise 95 In Example 3, find $x_1(t)$ explicitly from $x_1 = P(x_0)$ with $x_0(t) = 0$ for all t in $[0, 1]$. Find an upper bound on $\|x_1 - x\|$ where x is the solution in $S_{0.15}$. Show that $\|x_4 - x\| \leqslant 0.0000651$.

Exercise 96 Show that the problem in Example 3 has exactly two solutions. [Hint: show that any solution to the problem $x''(t) = e^{-x(t)}$, $x(0) = 0$ satisfies $e^{-x(t)} + x'(t)^2/2 = 1 + x'(0)^2/2$.]

CHAPTER 16

Fréchet derivatives

For nonlinear operator equations, the most efficient approach to the design of a successful computational method is often *local linearization*. This is the idea of 'Newton's method'. To apply the idea, we first need a concept of derivative for nonlinear operators. This is the subject of this chapter. The Fréchet derivative of a nonlinear operator is a generalization of the derivative of a real-valued function. It enables us to find a linear approximation to a nonlinear operator in a neighborhood of some given point (a local linearization). If we then replace the given nonlinear operator equation by its local linearization, we can use the methods for linear operator equations to find an approximate solution to the nonlinear operator equation. Regarding this as a point in the appropriate function space, we can again find a local linearization of the original nonlinear operator equation in a neighborhood of this new point. We can iterate this process, and it is a generalization of Newton's method for solving an equation in a single real variable.

In the next chapter, we will discuss Newton's method itself for nonlinear operator equations. In the present chapter, we first build the necessary tools we will need to construct the required local linearizations, namely Fréchet derivatives, which are bounded linear operators.

The main advantage of Newton-type methods over many other iterative methods is quadratic convergence. This means that, once we get close to a solution, the error (measured in an appropriate norm or metric) is squared at each iteration. Thus, Newton's method is rapidly convergent in practice, *if* we can find a good first guess at the solution. In a later chapter, we will discuss an approach to finding a good first approximation for Newton's method, based on the mathematical concept of a 'homotopy'.

Let $P : S \subseteq B_1 \to B_2$ be an operator mapping a subset S of a Banach space B_1 into a Banach space B_2. Let x_0 be an element of B_1 such that S contains a neighborhood of x_0. Then P is said to be *Fréchet differentiable* at x_0 if there is a continuous *linear* operator. $L : B_1 \to B_2$ such that: for every $\epsilon > 0$ there is a $\delta(\epsilon) > 0$ for which $\|P(x) - \{P(x_0) + L(x - x_0)\}\|_{B_2} \leqslant \epsilon \, \|x - x_0\|_{B_1}$ whenever $x \in N_{\delta(\epsilon)}(x_0) = \{x : \|x - x_0\|_{B_1} \leqslant \delta(\epsilon)\}$.

We can express this in the alternative form $P(x) = P(x_0) + L(x - x_0) + G(x, x_0)$ where $G : B_1 \to B_2$ defined by $G(x, x_0) = P(x) - P(x_0) - L(x, x_0)$ satisfies

$$\lim_{\|x - x_0\| \to 0} \left\{ \frac{\|G(x, x_0)\|_{B_2}}{\|x - x_0\|_{B_1}} \right\} = 0 .$$

Put still another way, this expresses the fact that $P(x) - P(x_0)$ is *locally linear* at x_0. In a sufficiently small neighborhood of x_0, $P(x) - P(x_0)$ can be approximated arbitrarily closely (in the Banach space norms) by image $L(x - x_0)$ of the linear operator L.

If such a continuous (bound) *linear operator* L exists for a particular x_0 in B_1 we denote it by $P'(x_0)$, *the Fréchet derivative of* P *at* x_0. Thus, if P is Fréchet differentiable at x_0 we can write $P(x) = P(x_0) + P'(x_0)(x - x_0) + G(x, x_0)$ where $G(x, x_0)$ satisfies the condition given above.

A few examples of Fréchet derivatives may be helpful.

Example 1

Denote the real line by R and let $[a, b]$ be an interval in R. Now R is a Banach space with norm $\|x\| = |x|$ for x in R.

Let $f : [a, b] \subseteq R \to R$ be a differentiable real valued function with $f(x)$ defined f for x in $[a, b]$, and let x_0 be in $[a, b]$; then the Fréchet derivative of f at x_0 is the ordinary derivative of f at x_0, $f'(x_0)$. Since

$$f'(x_0) = \lim_{|x - x_0| \to 0} \frac{f(x) - f(x_0)}{x - x_0} ,$$

it follows that $f(x) = f(x_0) + f'(x_0)(x - x_0) + G(x, x_0)$,

with
$$\lim_{|x - x_0| - 0} \frac{|G(x, x_0)|}{|x - x_0|} = 0.$$

Note that, as a real number, $f'(x_0)$ does represent a continuous linear mapping $f'(x_0) : R \to R$. In fact, $f'(x_0)(y)$ is defined by *scalar multiplication*: $f'(x_0)(y) = f'(x_0)y$.

Example 2

The n-dimensional Euclidean space, E^n, with norm

$$\|x\| = \left(\sum_{i=1}^{n} x_i^2 \right)^{1/2}$$

is also a Banach space. Let $P : S \subseteq E^n \to E^n$ be a mapping defined on a subset S of E^n which contains a neighborhood

$$N_{\bar{\delta}}(x_0) = \{x : \|x - x_0\| \leqslant \bar{\delta}\}$$

of a point x_0 in S.

It is not hard to show that $P'(x_0)$, if it exists, is the Jacobian matrix with elements

$$(P'(x_0))_{ij} = \left.\frac{\partial P_i}{\partial x_j}\right|_{x_0}, \quad i, j = 1, 2, \ldots, n.$$

A sufficient condition for the existence of $P'(x_0)$ is the continuity at x_0 of the partial derivatives. In this case, we have

$$P_i(x) = P_i(x_0) + \sum_{j=1}^{n} \left.\frac{\partial P_i}{\partial x_j}\right|_{x_0} (x - x_0)_j + G_i(x, x_0), i = 1, 2, \ldots, n.$$

The limit $DP(x_0)(y) = \lim_{a \to 0} \dfrac{P(x_0 + ay) - P(x_0)}{a}$,

if it exists, is called the *Gâteaux derivative of P at x_0 in the direction y*. (See: R. A. Tapia, 'The differentiation and integration of nonlinear operators,' in *Nonlinear functional analysis and applications* (L. B. Rall, ed.), pp. 45–101.)

For $P : S \subseteq E^n \to E^n$ with S containing a neighborhood $N_\delta(x_0)$ of x_0, suppose the partial derivatives $J_{ij}(x_0) = \left.\dfrac{\partial P_i}{\partial x_j}\right|_{x_0}$ exist in $N_\delta(x_0)$ and are continuous at X_0 for $i, j = 1, 2, \ldots, n$. Let $J(x_0)$ be the Jacobian matrix with elements $J_{ij}(x_0)$. We have the following.

Theorem The Fréchet derivative of P at x_0, $P'(x_0)$ exists, and $P'(x_0) = J(x_0)$ if and only if $DP(x_0)(y)$ exists and $DP(x_0)(y) = J(x_0)y$ for every y in E^n.

Proof The condition $DP(x_0)(y) = J(x_0)y$ for all y in E^n can be interpreted geometrically as saying that for each of the surfaces $P_i(x_1, x_2, \ldots, x_n)$, in E^{n+1}, $(i = 1, 2, \ldots, n)$, the tangent lines to the surface P_i through the point x_0 exist for all directions y in E^n, and that all these tangent lines lie in an n-dimensional tangent plane spanned by the tangent lines for the n directions corresponding to the coordinate axes in E^n: $(1, 0, 0, \ldots, 0)$, $(0, 1, 0, 0, \ldots)$, $\ldots (0, 0, \ldots, 1)$. More precisely, the condition requires $DP(x_0)(y)$ to be *linear* in y; exactly what is needed for the existence of $P'(x_0)$.

An analytic proof is as follows. If $P'(x_0) = J(x_0)$, then

$$\lim_{\|ay\| \to 0} \frac{\|P(x_0 + ay) - P(x_0) - J(x_0)ay\|}{\|ay\|} = 0,$$

and it follows that, for each y in E^n with $\|y\| \neq 0$,

$$\lim_{a \to 0} \left\| \frac{P(x_0 + ay) - P(x_0)}{|a|} - \frac{aJ(x_0)y}{|a|} \right\| = 0.$$

For $a < 0$, $|a| = -a$ and $\dfrac{-a}{|a|} = 1$; for $a > 0$, $|a| = a$ and $\dfrac{-a}{|a|} = -1$. In either case, we have

$$\lim_{a \to 0} \left\| \frac{P(x_0 + ay) - P(x_0)}{a} - J(x_0)y \right\| = 0.$$

Therefore, $DP(x_0)(y) = J(x_0)y$.

Conversely, if

$$DP(x_0)(y) = \lim_{a \to 0} \frac{P(x_0 + ay) - P(x_0)}{a} = J(x_0)y$$

for every y in E^n, then

$$P_i(x_0 + y) - P_i(x_0) = \sum_{j=1}^{n} \left(\frac{\partial P_i}{\partial y_j} \bigg|_{x_0 + \theta_i y} \right) y_j$$

for some $0 < \theta_i < 1, i = 1, 2, \ldots, n$.
Using the norm $\|y\| = \max\limits_i |y_i|$ in E^n, we have

$$\|P(x_0 + y) - P(x_0) - J(x_0)y\| \leqslant b(y)\, \|y\|$$

where $$b(y) = \max_i \sum_{j=1}^{n} \left| \left[\frac{\partial P}{\delta y_j} \right]_{x_0 + \theta_i y} - \left[\frac{\partial P_i}{\partial x_j} \right]_{x_0} \right|.$$

Now from the continuity of $\dfrac{\partial P_i}{\partial y_j}$ at x_0 it follows that given $\epsilon > 0$ there is a $\delta(\epsilon) > 0$ such that $b(y) \leqslant \epsilon$ whenever $\|y\| \leqslant \delta(\epsilon)$.

The Jacobian matrix $J(x_0)$ *is* a bounded linear transformation on E^n, so $P'(x_0)$ exists and $P'(x_0) = J(x_0)$. This completes the proof.

As a special application of the theorem just proved, consider a mapping $f: E^2 \to E^2$ viewed as a complex valued function of a complex variable, identifying a point (x, y) in E^2 with the complex number $x + iy$. Denote the components of f by u and v. Then $f(x, y) = u(x, y) + iv(x, y)$ corresponds to

$$f(x, y) = \begin{pmatrix} u(x, y) \\ v(x, y) \end{pmatrix}$$

in column vector notation for an image of the mapping f. Applying the theorem just proved, the Fréchet derivative $f'(z_0)$ exists at $z_0 = x_0 + iy_0$ and is expressible as the Jacobian matrix $f'(z_0) = J(z_0)$ given by

$$J(z_0) = \begin{pmatrix} u_x & u_y \\ v_x & v_y \end{pmatrix}$$

where $\quad u_x = \left.\dfrac{\partial u}{\partial x}\right|_{z_0}, \; u_y = \left.\dfrac{\partial u}{\partial y}\right|_{z_0}, \; v_x = \left.\dfrac{\partial v}{\partial x}\right|_{z_0},$

and $\quad v_y = \left.\dfrac{\partial v}{\partial y}\right|_{z_0}$ provided that for every w in E^2 the limit

$$\mathrm{D}f(z_0)(w) = \lim_{a \to 0} \frac{f(z_0 + aw) - f(z_0)}{a}$$

exists and $\mathrm{D}f(z_0)w = J(z_0)w$.

Now suppose the limit does exist for every w in E^2 and that $\mathrm{D}f(z_0)w = J(z_0)w$. Then, regarding w and $f'(z_0)w$ as complex numbers, we have

$$\begin{aligned} f'(z_0)w = J(z_0)w &= \begin{pmatrix} u_x w_1 + u_y w_2 \\ v_x w_1 + v_y w_2 \end{pmatrix} \\ &= (u_x w_1 + u_y w_2) + i(v_x w_1 + v_y w_2) \\ &= \frac{(u_x w_1 + u_y w_2) + i(v_x w_1 + v_y w_2)}{w} w \,; \end{aligned}$$

so that $f'(z_0)$ acting on w can be represented, via *complex multiplication* with $w = w_1 + iw_2$, by the complex number

$$f'(z_0) = \frac{u_x w_1^2 + v_y w_2^2 + (u_y + v_x)w_1 w_2 + i\{v_x w_1^2 - u_y w_2^2 + (v_y - u_x)w_1 w_2\}}{w_1^2 + w_2^2}.$$

Since $f'(z_0)$ is independent of w, we must have (the *Cauchy–Riemann equations*) $u_x = v_y$ and $u_y = -v_x$, then $f'(z_0) = u_x + iv_x = v_y - iu_y$. Thus,

$$f'(z_0) = \begin{pmatrix} u_x & -v_x \\ v_x & u_x \end{pmatrix} = u_x I + v_x J$$

where $\quad I = \begin{pmatrix} 1 & 0 \\ 0 & 1 \end{pmatrix}$ and $J = \begin{pmatrix} 0 & -1 \\ 1 & 0 \end{pmatrix}$.

This matrix representation of $f'(z_0)$ can be seen to be algebraically equivalent to the complex number representation $f'(z_0) = u_x + iv_x$ by means of an isomorphism between the complex number field and 2×2 matrices of the form $aI + bJ$. In fact, we have $I^2 = 1$ and $J^2 = -1$ so that, putting $I \sim 1$ and $J \sim i$ we have $aI + bJ \sim a + bi$ for any real numbers a and b. For instance, the product of any two such matrices yields $(a_1 I + b_1 J)(a_2 I + b_2 J) = (a_1 a_2 - b_1 b_2)I + (a_1 b_2$

$+ a_2 b_1)J$ corresponding to the complex number $(a_1 + b_1 i)(a_2 + b_2 i) = (a_1 a_2 - b_1 b_2) + (a_1 b_2 + a_2 b_1)i$.

Example 3

Let $B_1 = B_2 = C[0, 1]$ with norm $\|x\| = \max_{t \in [0,1]} |x(t)|$ and let $P : S \subseteq C[0, 1] \to C[0, 1]$ be defined by an integral operator of the form

$$P(x)(t) = g(t) + \int_0^1 f(t, s, x(s))\,\mathrm{d}s$$

where g is in $C[0, 1]$ and $S = \{x : \|x\| < b\}$ for some $b > 0$.

For x_0 in S we can write

$$P(x)(t) - P(x_0)(t) = \int_0^1 \{f(t, s, x(s)) - f(t, s, x_0(s))\}\,\mathrm{d}s$$

and so, if $\left.\frac{\partial f}{\partial x}\right|_{t,s,x_0(s)}$ is continuous for t and s in $[0, 1]$ we have

$$P(x)(t) - P(x_0)(t) = \int_0^1 \left\{ \left.\frac{\partial f}{\partial x}\right|_{t,s,x_0(s)} (x(s) - x_0(s)) \right\} \mathrm{d}x + G(x, x_0),$$

and the Fréchet derivative of P at x_0 is the continuous linear differential operator $P'(x_0)$ on $C[0, 1]$ defined by

$$(P'(x_0)u)(t) = \int_0^1 \frac{\partial f}{\partial x}(t, s, x_0(s))u(s)\,\mathrm{d}x.$$

Example 4

If $L : B_1 \to B_2$ is a continuous *linear operator*, then the Fréchet derivative of L at x_0 is the operator L itself. $L'(x_0) \equiv L$ for all x_0 in B_1. To see this, we write $Lx - Lx_0 = L'(x_0)(x - x_0) + G(x, x_0)$ and, since L is linear, we have $Lx - Lx_0 \equiv L'(x_0)(x - x_0)$ if we choose $L'(x_0) \equiv L$; and $G \equiv 0$.

The definition and examples of Fréchet derivatives may raise the question of uniqueness.

Suppose L_1 and L_2 are linear operators both satisfying

$$P(x) = P(x_0) + L_1(x - x_0) + G_1(x, x_0)$$

$$P(x) = P(x_0) + L_2(x - x_0) + G_2(x, x_0)$$

where G_1 and G_2 both satisfy the property making L_1 and L_2 Fréchet derivatives of P at x_0.

Now $L_1 - L_2$ is a linear operator on B_1 defined by $(L_1 - L_2)u = L_1 u - L_2 u$ and $(L_1 - L_2)(x - x_0) = G_2(x, x_0) - G_1(x, x_0)$ for all x in B_1. Thus

$$\|L_1 - L_2\| = 0; \quad \text{hence } L_1 = L_2 .$$

Thus, when the Fréchet derivative exists, it is unique.

Exercise 97 Show that if $P'(x_0)$ exists, then P is continuous at x_0.

Let B_1 and B_2 be Banach spaces. Suppose that $P'(x_0) : B_1 \to B_2$ is *one-one, onto*; then the linear operator equation $P'(x_0)x = y$ has a unique solution in B_1, and $[P'(x_0)]^{-1}$ exists as a bounded one-one linear operator from B_2 onto B_1 and $x = [P'(x_0)]^{-1} y$, for every y in B_2, (Kolmogoroff and Fomin, *Functional analysis*, Vol. 1, pp. 98–101).

By a δ *neighborhood* of x_0 in B_1 we will mean a set $N_\delta(x_0) = \{x : x \in B_1, \|x - x_0\| \leqslant \delta\}$ for some $\delta > 0$.

Exercise 98

a) Find the Fréchet derivative $P'(x_0)$ for the operator $P: C_0^{(2)}[0, 1] \to C[0, 1]$ defined by $P(u) = u'' - e^{+u}$. It can be shown (P. Henrici, *Discrete variable methods* . . . , 1962, p. 347.) that a boundary value problem of the form $y'' = f(t, y), y(a) = A, y(b) = B$ with

$$\frac{\partial f}{\partial y}(t, y) \geqslant 0 \quad \text{for } a \leqslant t \leqslant b, \; -\infty \leqslant y \leqslant \infty$$

has a unique solution. Show that this implies that $P'(x_0)$ has an inverse for every x_0 in $C_0^{(2)}[0, 1]$.

b) For $P: C^{(2)}[0, 1] \to C[0, 1]$ (no boundary conditions) given by $P(u) = u'' - e^{-u}$, show that $P'(x_0)$ does not have an inverse at $x_0(t) \equiv - \ln \pi^2$ where

$$\|u\|_{C^{(2)}[0,1]} = \max_{t \in [0,1]} (|u(t)| + |u''(t)|) .$$

Theorem Let P be an operator $P: S \subseteq B_1 \to B_2$ with S containing a neighborhood $N_\delta(x_0)$. If $[P'(y)]^{-1}$ exists and is uniformly bounded in $N_\delta(x_0)$ then P is one-one in some neighborhood of x_0.

Proof If $P(x) = P(y)$ for some x, y in $N_\delta(x_0)$ then $P'(y)(x - y) = -G(x, y)$, where G is such that given $\epsilon > 0$ there is a $\delta(\epsilon) > 0$ for which $\|x - y\| \leqslant \delta(\epsilon)$ implies $\|G(x, y)\| \leqslant \epsilon \|x - y\|$. From $x - y = -[P'(y)]^{-1} G(x, y)$ it follows that $\|x - y\| \leqslant \|[P'(y)]^{-1}\| \, \epsilon \, \|x - y\|$ whenever $\|x - y\| \leqslant \delta(\epsilon)$.

Thus, one of the following is true:

1) $\|x - y\| > \delta(\epsilon)$, or
2) $x = y$, or
3) $x \neq y$ but $\|x - y\| \leqslant \delta(\epsilon)$ and $\|[P'(y)]^{-1}\| \geqslant 1/\epsilon$

Now since $\|[P'(y)]^{-1}\| \leqslant M$ for y in $N_\delta(x_0)$ we can rule out the third possibility for $\epsilon < 1/M$, that is for $\|x-y\| < \delta(1/M)$. Thus, for x, y in $N_{\delta'}(x_0)$ with $\delta' < \frac{1}{2}\delta(1/M)$ we have $\|x-y\| \leqslant \|x-x_0\| + \|y-y_0\| \leqslant 2\delta' < \delta(1/M)$ and it follows from $P(x) = P(y)$ that $x = y$. P is therefore one-one on $N_{\delta'}(x_0)$.

Corollary If x_0 is a zero of P and the inverse of the Fréchet derivative of P exists and is uniformly bounded on a neighborhood $N_\delta(x_0)$ included in the domain of P, then x_0 is a *simple* zero of P_i, that is P is one-one on $N_\delta(x_0)$.

Exercise 99 For an operator $P: C[0,1] \to C[0,1]$ of the form

$$P(u)(t) = u(t) - g(t) - \int_0^1 f(t,s,u(s))\,ds,$$

with g in $C[0,1]$, derive sufficient conditions on f to insure that $[P'(x_0)]^{-1}$ exists and is uniformly bounded on a $\delta > 0$ neighborhood of 0 in $C[0,1]$. Show that P can have at most one zero in $N_{\bar\delta}(0)$ under those conditions for some $0 < \bar\delta$.

Let $L(B_1, B_2)$ be the Banach space of bounded linear operators from B_1 to B_2 with norm

$$\|L\| = \sup_{\substack{x \in B_1 \\ x \neq 0}} \frac{\|Lx\|_{B_2}}{\|x\|_{B_1}}.$$

If $P: S \subseteq B_1 \to B_2$ has a Fréchet derivative $P'(x)$ at each x in $N_\delta(x_0)$ contained in S then we may consider whether P' as a mapping

$$P': N_\delta(x_0) \subseteq B_1 \to L(B_1, B_2)$$

itself has a Fréchet derivative at x_0. If it does, we will denote it by

$$P''(x_0): B_1 \to L(B_1, B_2),$$

the *second Fréchet derivative of* P *at* x_0. Then $P''(x_0)$ is a bounded linear operator on B_1 whose image at u in B_1 is a bounded linear operator $P''(x_0)u$ from B_1 into B_2, and $P'(x) = P'(x_0) + P''(x_0)(x - x_0) + G(x, x_0)$ where $G(x, x_0)$ satisfies

$$\lim_{\|x-x_0\| \to 0} \frac{\|G(x,x_0)\|_{L(B_1,B_2)}}{\|x-x_0\|_{B_1}} = 0.$$

If $P: B_1 \to B_2$ is a continuous *linear* operator then $P'(x_0) \equiv P$ for all x_0 in B_1 (see Example 4). In this case, $P''(x_0) \equiv 0$, the operator which maps every u in B_1 into the linear operator 0 in $L(B_1, B_2)$. To see this, we note that, in this case, $P'(x) \equiv P'(x_0) = P$, so that $P''(x_0) \equiv 0$ and $G(x, x_0) \equiv 0$.

For an operator $P: E^n \to E^n$, $P''(x_0)$, if it exists, is the *tensor* with components $(P''(x_0))_{ijk}$ such that $P'(x) = P'(x_0) + P''(x_0)(x - x_0) + \ldots$, so that

$$(P'(x))_{ij} = (P'(x_0))_{ij} + \sum_{k=1}^{n} (P''(x_0))_{ijk} (x - x_0)_k + \ldots$$

In fact $$(P''(x_0))_{ijk} = \left.\frac{\partial^2 P_i}{\partial x_j \, \partial x_k}\right|_{x_0};$$

so that for the first few terms of the Taylor series, for instance, we have

$$P_i(x) = P_i(x_0) + \sum_{j=1}^{n} \left.\frac{\partial P_i}{\partial x_j}\right|_{x_0} (x - x_0)_j$$
$$+ \frac{1}{2} \sum_{j,k=1}^{n} \left.\frac{\partial^2 P_i}{\partial x_j \, \partial x_k}\right|_{x_0} (x - x_0)_k (x - x_0)_j + \ldots$$

Note that $P''(x_0)u$ is a linear operator on E^n with

$$(P''(x_0)u)_{ij} = \sum_{k=1}^{n} \left.\frac{\partial^2 P_i}{\partial x_j \, \partial x_k}\right|_{x_0} (x - x_0)_k, \quad i, j = 1, 2, \ldots, n.$$

Exercise 100

a) If $[P'(x_0)u](t) = u''(t) + f(x_0(t))u(t)$, then show that $P''(x_0)$ is represented formally by $[[P''(x_0)u]\, v](t) = f'(x_0(t))u(t)\, v(t)$.

b) If $P(u)(t) = u(t) - g(t) - \int_0^1 f(t, s, u(s))\,ds$, find formal representations of $P'(x_0)$ and $P''(x_0)$.

c) If $P(u)(t) = u''(t) - e^{-u(t)}$, find upper bounds on $\|P'(x_0)\|$ and $\|P''(x_0)\|$

where $P'(x_0): C_0^{(2)}[0, 1] \to C[0, 1]$

and $P''(x_0): C_0^{(2)}[0, 1] \to L(C_0^{(2)}[0, 1], C[0, 1])$.

(Hint: for u in $C_0^{(2)}[0, 1]$ we have $u(0) = u(1) = 0$ so that

$$u(t) = u'(0)t + \frac{1}{2} u''(\xi_t)t^2$$

for some ξ_t in $[0, 1]$. Thus

$$u(1) = 0 = u'(0) + \frac{1}{2} u''(\xi_1);$$

and so $\|u\|_{C[0,1]} \leqslant \|u\|_{C_0^{(2)}[0,1]}$.)

d) If $P(u) = u_t - K(u)u_{xx}$, where

$$u_t = \frac{\partial u(t, x)}{\partial x} \quad \text{and} \quad u_{xx} = \frac{\partial^2 u(t, x)}{\partial x^2},$$

find $P'(u_0)$ and $P''(u_0)$ formally.

e) If $P(u)(t) = u''(t) - 6u^2(t) - 5t$, express $P'(u_0)$ and $P''(u_0)$ formally.

Higher Fréchet derivatives are defined recursively, for $n = 2, 3, \ldots$, by $P^{(n)}(x_0) : B_1 \to L(B_1^{n-1}, B_2)$ such that

$$\lim_{\|x - x_0\| \to 0} \frac{\|P^{(n-1)}(x) - P^{(n-1)}(x_0) - P^{(n)}(x_0)(x - x_0)\|}{\|x - x_0\|} = 0$$

where $L(B_1^1, B_2) = L(B_1, B_2)$ and $L(B_1^n, B_2) = L(B_1, L(B_1^{n-1}, B_2))$, (see L. B. Rall, *Computational solution of nonlinear operator equations*, Wiley, 1969; reprinted by Krieger, Huntington, N.Y., 1979).

We can also view $P^{(n)}(x_0)$ as a mapping from B_1^n (the Cartesian product of B_1 with itself n times; that is, n-tuples $(x_1, x_2, \ldots, x_n)$ of elements x_i in B_1) to B_2. We can write the image of $P^{(n)}(x_0)$ at such a point as $P^{(n)}(x_0)x_1x_2 \ldots x_n$. This will be linear in each of the variables $x_1, x_2, \ldots, x_n$ and *symmetric*; that is, $P^{(n)}(x_0)x_1x_2 \ldots x_i \ldots x_j \ldots x_n = P^{(n)}(x_0)x_1 \ldots x_j \ldots x_i \ldots x_n$ for any $1 \leqslant i < j \leqslant n$. If $x_1 = x_2 = \ldots = x_n = x$, we write $P^{(n)}(x_0)xx \ldots x = P^{(n)}(x_0)x^n$.

We denote by $\overline{xy}$ the line segment $\overline{xy} = \{\theta x + (1 - \theta)y : 0 \leqslant \theta \leqslant 1\} \subseteq B_1$ when x and y are elements of B_1. If $P : B_1 \to B_2$ maps the Banach space B_1 into B_2, then P maps the line segment $\overline{xy}$ in B_1 into the *arc* $P(\overline{xy}) = \{P(\theta x + (1 - \theta)y : 0 \leqslant \theta \leqslant 1\}$ in B_2.

We define the integral

$$\int_0^1 P(\theta x + (1 - \theta)y)\mathrm{d}\theta = \lim_{n \to \infty} \sum_{i=1}^{n} P\left(\frac{i}{n}x + \left(1 - \frac{i}{n}\right)y\right)\frac{1}{n}$$

when the limit exists. (When P is not continuous, then arbitrary partitions of $0 \leqslant \theta \leqslant 1$ should be considered in the definition of the integral. For continuous P, the uniform partition will suffice.)

If P is continuously differentiable ($P'(x_0)$ continuous in x_0), then

$$P(x) = P(y) + \int_0^1 P'(\theta x + (1 - \theta)y)(x - y)\mathrm{d}\theta$$

In case P is only defined and continuous on a *subset* S of B_1, we can still perform the integration provided that S has a *convex* subset containing x and y (for instance a neighborhood $N_\delta(x_0)$). (Recall that a convex set in a Banach space is one which contains the line segment joining any two of its elements.)

It follows that, whenever P' is continuous on $\overline{xy}$, we have

$$\|P(x) - P(y)\| \leqslant \sup_{\bar{x} \in \overline{xy}} \|P'(\bar{x})\| \, \|x - y\| .$$

If $P: S \subseteq B_1 \to B_2$ is n times continuously differentiable on $N_\delta(x_0) \subseteq S$, then Taylor's theorem holds. (See: Rall, *loc. cit.*, p. 124) For any x in $N_\delta(x_0)$ we have

$$P(x) = P(x_0) + \sum_{i=1}^{n-1} \frac{P^{(i)}(x_0)}{i!} (x - x_0)^i + R_n(x, x_0)$$

where $$R_n(x, x_0) = \int_0^1 \frac{P^{(n)}(\theta x + (1-\theta)x_0)}{n!} (x - x_0)^n \, n(1-\theta)^{n-1} \, d\theta .$$

For $n = 2$, we have

$$\|P(x) - P(x_0) - P'(x_0)(x - x_0)\| \leqslant \sup_{\bar{x} \in \overline{x_0 x}} \left\| \frac{P^{(2)}(\bar{x})}{2} \right\| \|x - x_0\|^2$$

whenever x is in $N_\delta(x_0)$.

Exercise 101 Show that if $P^{(2)}$ is continuous and uniformly bounded by $\|P^{(2)}(x)\| \leqslant K$ for x in $N_\delta(x_0)$, and if $[P'(x_0)]^{-1}$ exists, then $[P'(x)]^{-1}$ exists and is uniformly bounded on $N_{\delta'}(x_0)$ for any

$$\delta' < \frac{1}{K \|[P'(x_0)]^{-1}\|} .$$

(Hint: Use the Taylor expansion with P' in place of P.)

CHAPTER 17

Newton's method in Banach spaces

In this chapter, we consider Newton's method in Banach spaces, and discuss the now classical theorem of L. V. Kantorovich concerning sufficient conditions for the convergence of the method. The method can be applied to a finite system of nonlinear equations, to integral equations, and to initial value and boundary value problems in differential equations.

The methods of interval analysis, see for example Chapter 14, can be helpful in bounding the norms of operators which occur in connection with Newton's method in Banach spaces. Alternatively, there are interval versions of Newton's method which proceed entirely with interval valued functions, or combine interval methods with the use of semi-norms. More detailed reference to such methods is given at the end of this chapter.

Let X_1, X_2 be Banach spaces and let $P: S \subseteq X_1 \to X_2$ be an operator whose domain S contains a neighborhood $N_\delta(x_0)$ in which the second Fréchet derivative exists, is continuous, and is uniformly bounded $\|P^{(2)}(y)\| \leqslant K$ for y in $N_\delta(x_0)$. Suppose further that $[P'(x_0)]^{-1}$ exists and $\|[P'(x_0)]^{-1}\| \leqslant B_0$, then we have

$$\|P(y) - P(x_0) - P'(x_0)(y - x_0)\| \leqslant \frac{K}{2} \|y - x_0\|^2$$

for y in $N_\delta(x_0)$. If P has a zero in $N_\delta(x_0)$, say $P(x) = 0$, then putting $y = x$, we obtain

$$\|-P(x_0) - P'(x_0)(x - x_0)\| \leqslant \frac{K}{2} \|x - x_0\|^2 \text{; so}$$

$$\|[P'(x_0)]^{-1} (P(x_0) + P'(x_0)(x - x_0))\| \leqslant \frac{B_0 K}{2} \|x - x_0\|^2 \text{, and}$$

$$\|x - \{x_0 - [P'(x_0)]^{-1} P(x_0)\}\| \leqslant \frac{B_0 K}{2} \|x - x_0\|^2 .$$

This inequality exhibits the *error squaring* property of *Newton's iterative method* for an approximate solution of the *nonlinear* equation $P(x) = 0$. The successive iterates $x_0, x_1, x_2, \ldots$ are defined recursively by

$$x_k = x_{k-1} - [P'(x_{k-1})]^{-1} P(x_{k-1})$$

so long as $[P'(x_{k-1})]^{-1}$ and $P(x_{k-1})$ exist; and x_k can be found from x_{k-1} by solving the *linear* operator equation (see Chapters 11–13 on methods for linear operator equations)

$$P'(x_{k-1})\,\Delta_k = -P(x_{k-1})$$

$$x_k = x_{k-1} + \Delta_k\,.$$

So long as x_{k-1}, and x_k remain in $N_\delta(x_0)$ for $k = 1, 2, \ldots$ we will have, provided that, for all k, $[P'(x_{k-1})]^{-1}$ exists and $\|[P'(x_{k-1})]^{-1}\| \leqslant B$, the result

$$\|x - x_k\| \leqslant \frac{BK}{2} \|x - x_{k-1}\|^2\,.$$

The *Kantorovich theorem* (Rall, see ref. in Chapter 16, pp. 133–144) provides sufficient conditions for the convergence of the Newton iterates to a solution of the operator equation $P(x) = 0$. In our notation these conditions are

$$\delta \geqslant \frac{1 - \sqrt{1 - 2h_0}}{h_0}\,\eta_0$$

$$h_0 = B_0 \eta_0 K \leqslant \frac{1}{2}$$

where $\quad \eta_0 \geqslant \|x_1 - x_0\|.$

For any particular numbers B_0 and K it will be possible to satisfy the above conditions if $\|x_1 - x_0\|$ is sufficiently small; this will be the case if $B_0\|P(x_0)\|$ is small enough, since $\|x_1 - x_0\| \leqslant B_0\,\|P(x_0)\|$.

Under the conditions given, the Newton sequence *is* defined (that is, $[P'(x_{k-1})]^{-1}$, $P(x_{k-1})$ do exist for $k = 0, 1, 2, \ldots$) and converges to a solution of the equation $P(x) = 0$.

The existence of $[P'(x_k)]^{-1}$ follows from that of $[P'(x_{k-1})]^{-1}$, provided that (for instance) $\|[P'(x_{k-1})]^{-1}\|\,K\,\|x_k - x_{k-1}\| < 1$.

If we write $P'(x_k) = P'(x_{k-1})\left\{I + [P'(x_{k-1})]^{-1}\,(P'(x_k) - P'(x_{k-1})\right\}$ then $[P'(x_k)]^{-1}$ can be seen to exist if $\|[P'(x_{k-1})]^{-1}\,(P'(x_k) - P'(x_{k-1}))\| < 1$ and would be representable by the Neumann series as a special case of $[I + E]^{-1} = I - E + E^2 - E^3 + \ldots$ with $\|E\| < 1$. The inequality above will be true if the previous one is true, making use of $\|P'(x_k) - P'(x_{k-1})\| \leqslant K\,\|x_k - x_{k-1}\|$.

A proof of the Kantorovich theorem which proceeds by an inductive argument is given by Rall (see especially Rall: *loc.cit.*, pp. 135–138).

The error-squaring property is computationally especially significant since it leads to very rapid convergence once the solution is approach fairly closely.

From the uniform boundedness of the second Fréchet derivative assumed in the conditions for the Kantorovich theorem, it follows that

$$\|[P'(y)]^{-1} \leqslant \frac{B_0}{1 - B_0 K\delta} \text{ for } y \text{ in } N_\delta(x_0),$$

provided that $B_0 K\delta < 1$. Under these conditions, it follows from the theorem and corollary in Chapter 16 that P has at most one zero in $N_\delta(x_0)$ and if there is one it is simple. The inequality above follows from $[P'(y)]^{-1} = [I + [P'(x_0)]^{-1} (P'(y) - P'(x_0))]^{-1} [P'(x_0)]^{-1}$.

Suppose now that P has a zero x^* and that $[P'(x^*)]^{-1}$ exists and $\|P^{(2)}(y)\| \leqslant K$ for y in $N_\delta(x^*)$. Then $[P'(y)]^{-1}$ exists for y in $N_\delta(x^*)$ and

$$\|[P'(y)]^{-1}\| \leqslant \frac{B^*}{1 - KB^*\delta}$$

provided that $KB^*\delta < 1$ where $B^* = \|[P'(x^*)]^{-1}\|$.

Then x^* is a simple zero of P, and the Newton iteration function for P, $NP(y) = y - [P'(y)]^{-1}P(y)$, is defined for all y in $N_\delta(x^*)$.

We have $P(x^*) = 0$ and $NP(x^*) = x^*$ and

$$\begin{aligned} NP(y) - x^* &= y - x^* + [P'(y)]^{-1}(P(x^*) - P(y)) \\ &= y - x^* + [P'(y)]^{-1}(P'(y)(x^* - y) + G(x^*, y)) \\ &= [P'(y)]^{-1} G(x^*, y) \end{aligned}$$

where $$\|G(x^*, y)\| \leqslant \frac{K}{2} \|x^* - y\|^2 .$$

Thus $$\|NP(y) - x^*\| \leqslant \frac{B^*K\|x^* - y\|}{2(1 - B^*K\delta)} \|y - x^*\| .$$

If $$\frac{B^*K\delta}{2(1 - B^*K\delta)} \leqslant \theta < 1,$$

then NP maps $N_\delta(x^*)$ into (not necessarily onto) $N_{\theta\delta}(x^*)$, and the successive iterates $x_{k+1} = x_k - [P'(x_k)]^{-1} P(x_k)$ converged to x^* from any x_0 in $N_\delta(x^*)$; in fact, $\|x_k - x^*\| \leqslant \theta^k\delta$, $k = 0, 1, 2, \ldots$ Thus the Newton iterates converge to the solution x^* from any x_0 in $N_\delta(x^*)$ whenever $\theta < 1$. For this, we need $\delta(B^*K + 2\theta KB^*) \leqslant 2\theta$ or

$$\delta \leqslant \frac{2\theta}{B^*K(1 + 2\theta)} \text{ with } \theta < 1.$$

From this we can conclude that the Newton iterates converge to x^* from any x_0 such that

$$\|x_0 - x^*\| < \frac{2}{3B^*K}$$

where $B^* = \|[P'(x^*)]^{-1}\|$ and $\|P^{(2)}(y)\| \leqslant K$ for y in $N_\delta(x^*)$ where

$$\delta = \frac{2}{3B^*K}.$$

This is a slightly sharper result than given by Rall (*loc. cit.* Th. 26.1, p. 188), who finds by a different argument that convergence to x^* follows from any x_0 such that

$$\|x^* - x_0\| < \frac{1}{4B^*K}$$

where $\|P^{(2)}(x)\| \leqslant K$ for x in $N_\delta(x^*)$ with $\delta > \dfrac{2}{B^*K}$.

To illustrate the difference between the two results, consider the simple case of the solution $x^* = 0$ of $P(x) = 0$ for the polynomial $P(x) = x - x^3$. From $\|x_0 - x^*\| < 2/(3B^*K)$ we conclude that the Newton iterates converge from any x_0 such that

$$|x_0| < \frac{1}{3};$$

whereas, from $\|x_0 - x^*\| < 1/(4\ B^*K)$ we conclude: only that convergence follows for

$$|x_0| < \frac{1}{8\sqrt{3}}.$$

Actually, the Newton iterates converge to $x^* = 0$ in this example if and only if

$$|x_0| < \frac{1}{\sqrt{5}}.$$

It is not a simple matter by any means (even for polynomials, let alone operators) to find the exact regions of starting points from which the Newton iterates will converge to a particular solution of a nonlinear equation. (See Rall's interesting discussion of this on pp.185–188, *loc.cit.*).

A question of practical interest for computational applications of Newton's method in Banach spaces is how to find some safe starting point x_0 from which the Newton iterates will converge to a solution of a nonlinear operator equation. See Chapter 19.

To apply the Kantorovich theorem we need an x_0 in $C_0^{(2)}[0, 1]$ for which it is possible to satisfy the following inequalities:

$$B_0 \geqslant \|[P'(x_0)]^{-1}\|,$$

$$\eta_0 \geqslant \|x_1 - x_0\|.$$

$$K \geqslant \|P^{(2)}(y)\| \text{ for all } \|y - x_0\| \leqslant \delta,$$

where $$\delta \geqslant \frac{1 - \sqrt{1 - 2h_0}}{h_0} \eta_0,$$

and $$h_0 = B_0 \eta_0 K \leqslant \frac{1}{2}.$$

It may not always be easy to tell whether $[P'(x_0)]^{-1}$ exists or, if it does, to find a bound on its norm. We give now three examples of operators for which we can find a B_0 satisfying the first of the inequalities above. The other inequalities will be discussed as well in Example 3 below.

Example 1

The Fréchet derivative at x_0 of $P : E^n \to E^n$, is the Jacobian matrix

$$[P'(x_0)]_{ij} = \left. \frac{\partial P_i}{\partial x_j} \right|_{x_0}, \quad (i, j = 1, 2, \ldots, n)$$

provided that these partial derivatives exist and are continuous at x_0. Thus, if this Jacobian matrix is non-singular then we can take any upper bound on the norm of its inverse for B_0.

Example 2

The Fréchet derivative at x_0 of an integral operator $P : C[0, 1] \to C[0, 1]$ of the form

$$[P(u)](t) = u(t) - g(t) - \int_0^1 K(t, s) f(s, u(s)) \, ds$$

is expressible as

$$[P'(x_0)y](t) = y(t) - \int_0^1 K(t, s) \frac{\partial f}{\partial u}(s, x_0(s)) y(s) \, ds$$

provided that K is continuous on the unit square and $\dfrac{\partial f}{\partial u}(s, x_0(s))$ is continuous on $[0, 1]$. If we define the linear operator Q on $C[0, 1]$ by

$$[Qy](t) = \int_0^1 K(t, s) \frac{\partial f}{\partial u}(s, x_0(s))y(s)ds$$

then we have $P'(x_0)y = (I - Q)y$ or $P'(x_0) = I - Q$. Thus, if $\|Q\| = \sup_{\|y\|=1} \|Qy\| < 1$, we know that $[P'(x_0)]^{-1}$ exists and that

$$\|[P'(x_0)]^{-1}\| \leqslant \frac{1}{1 - \|Q\|}.$$

We will have $\|Q\| < 1$ if, for instance,

$$\|Q\| \leqslant \max_{t \in [0,1]} \int_0^1 |K(t, s)| \left| \frac{\partial f}{\partial u}(s, x_0(s)) \right| ds < 1;$$

and we can take $B_0 \geqslant \dfrac{1}{1 - \|Q\|}$.

Example 3

The Fréchet derivative at x_0 of a differential operator $P : C_0^{(2)}[0, 1] \to C[0, 1]$ of the form $[P(u)](t) = u''(t) - f(t, u(t))$ can be written as

$$[P'(x_0)y](t) = y''(t) - \frac{\partial f}{\partial u}(t, x_0(t))y(t)$$

provided that $\dfrac{\partial f}{\partial u}(t, x_0(t))$ is continuous in t for t in $[0, 1]$.

It can be shown that if the initial value problem $w_1''(t) - a(t)w_1(t) = 0$ with

$$w_1(0) = 0, w_1'(0) = 1, \ a(t) = \frac{\partial f}{\partial u}(t, x_0(t))$$

has the solution $w_1(t)$ for $0 \leqslant t \leqslant 1$, and if $w_1(1) \neq 0$, then the equation $[P'(x_0)y](t) = z(t)$ has a unique solution for every z in $C[0, 1]$. In this case $[P'(x_0)]^{-1} : C[0, 1] \to C_0^{(2)}[0, 1]$ can be expressed as

$$[P'(x_0)]^{-1} z(t) = \int_0^1 G(t, s)z(s)ds,$$

where $G(t, s)$ is the *Green's function* for the linear differential operator $P'(x_0)$ on $C_0^{(2)}[0, 1]$ (boundary conditions: $y(0) = y(1) = 0$ for all y in $C_0^{(2)}[0, 1]$) given by

$$G(t, s) = \begin{cases} \dfrac{1}{w_1(1)} w_1(t)w_2(s), & 0 \leqslant t \leqslant s \leqslant 1 \\ \dfrac{1}{w_1(1)} w_1(s)w_2(t), & 0 \leqslant s \leqslant t \leqslant 1 \end{cases}$$

where $w_2(t)$ is the solution to the initial value problem $w_2''(t) - a(t)w_2(t) = 0$

with $$w_2(1) = 0, w_2'(1) = 1, a(t) = \frac{\partial f}{\partial u}(t, x_0(t)).$$

In this case we will have, since

$[P'(x_0)]^{-1} : C[0, 1] \to C_0^{(2)}[0, 1]$, the upper bound

$$\|[P'(x_0)^{-1}\| = \sup_{\|z\|=1} \|[P'(x_0)]^{-1} z\|_{C_0^{(2)}[0,1]} \cdot$$

Thus, we can take $B_0 \geqslant \max_{t \in [0,1]} \left\{ \frac{d^2}{dt^2} \int_0^1 |G(t, s)| \, ds \right\}$

For a specific case of an example of this type consider $P(u) = u'' - e^{-u}$. To find a zero of P in $C_0^{(2)}[0, 1]$ using Newton's method we might try $x_0(t) = \frac{1}{2} t(t - 1)$. In any case we want an x_0 which will make $\|x_1 - x_0\| = \|[P'(x_0)]^{-1} P(x_0)\|$ small. With this x_0 we have $P(x_0)(t) = 1 - e^{(1/2)t(1-t)}$ and $\|P(x_0)\| = e^{(1/8)} - 1 < 0.134$; and we have

$$\|[P'(x_0)\| \leqslant \max_{t \in [0,1]} \frac{d^2}{dt^2} \int_0^1 |G(t, s)| \, ds$$

where G is given above with $w_1(t)$ satisfying, for $a(t) = -\exp(-x_0(t))$,

$$w_1''(t) + e^{(1/2)t(1-t)} w_1(t) = 0$$

$$w_1(0) = 0, w_1'(0) = 1$$

and $w_2(t)$ satisfying

$$w_2''(t) + e^{(1/2)t(1-t)} w_2(t) = 0$$

$$w_2(1) = 0, \; w_2'(1) = 1.$$

Now, assuming for the moment that $w_1(1) \neq 0$, we have

$$\int_0^1 |G(t, s)| \, ds = \frac{1}{w_1(1)} \left\{ |w_2(t)| \int_0^t |w_1(s)| ds \right.$$

$$\left. + |w_1(t)| \int_t^1 |w_2(s)| ds \right\},$$

and

$$\frac{d^2}{dt^2} \int_0^1 |G(t, s)| ds = \frac{1}{w_1(1)} \left\{ |w_2(t)|'' \int_0^t |w_1(s)| ds \right.$$

$$+ |w_1(t)|'' \int_t^1 |w_2(s)| \mathrm{d}s$$

$$+ |w_2(t)|' \, |w_1(t)| - |w_1(t)|' \, |w_2(t)| \Big\}$$

We can bound the expression above and hence $\|P'(x_0)^{-1}\|$ using interval methods as discussed in Chapter 14. To do this we can first rewrite the differential equations for w_1 and w_2 as integral equations. Integrating them, we obtain

$$w_1{}'(\bar{t}) = 1 + \int_0^{\bar{t}} \{-\mathrm{e}^{(1/2)s(1-s)} w_1(s)\} \, \mathrm{d}s$$

$$w_1(t) = t - \int_0^t \int_0^{\bar{t}} \mathrm{e}^{(1/2)s(1-s)} w_1(s) \mathrm{d}s \mathrm{d}\bar{t}$$

$$= t - \int_0^t \int_s^t \mathrm{e}^{(1/2)s(1-s)} w_1(s) \mathrm{d}\bar{t} \mathrm{d}s \, ,$$

so $$w_1(t) = t - \int_0^t (t-s) \, \mathrm{e}^{(1/2)s(1-s)} w_1(s) \mathrm{d}s \, .$$

Similarly, $$w_2(t) = t - 1 - \int_t^1 (s-t) \, \mathrm{e}^{(1/2)s(1-s)} w_2(s) \mathrm{d}s \, .$$

Now, using the techniques of interval analysis (Chapter 14) we find that

$$\mathrm{e}^{(1/2)s(1-s)} \subseteq [1, \mathrm{e}^{0.125}] \subseteq [1, 1.134] \text{ for all } s \text{ in } [0, 1] \, ;$$

and so we consider the interval operator related to the integral equation for w_1 given by

$$Q_1(X_0)(t) = t - \int_0^t (t-s)[1, 1.134] \, X_0(s) \mathrm{d}s.$$

For $X_0(t)$ of the form $X_0(t) = [a, 1] t, 0 < a < 1$ we have

$$Q_1(X_0)(t) = t - [a, 1.134] \int_0^t (t-s) s \mathrm{d}s$$

$$= t - \left[\frac{a}{6}, 0.189\right] t^3 \, ,$$

and we will have $Q_1(X_0)(t) \subseteq X_0(t)$ for all t in $[0, 1]$ provided that

$$t - \left[\frac{a}{6}, 0.189\right] t^3 \subseteq [a, 1] t, \text{ for all } t \text{ in } [0, 1] \, .$$

The relation above will be satisfied for $a > 0$ such that $1 - 0.189t^2 \geqslant a$, $(t \in [0, 1])$. Hence, we may take $a = 0.811$. Then we have $w_1(t) \subseteq t - [0.135, 0.189]t^3 \subseteq [0.811, 1]\, t$.

Note that the above inclusions give us a choice of two interval polynomials containing $w_1(t)$. In particular $w_1(1) \subseteq [0.811, 0.865]$.

Next, for an interval operator related to the integral equation for w_2, consider

$$Q_2(X_0)(t) = t - 1 - \int_t^1 (s - t)[1, 1.134]X_0(s)\mathrm{d}s\,.$$

For $X_0(t)$ of the form $X_0(t) = [a, 1](t - 1), 0 < a < 1$ we have

$$Q_2(X_0)(t) = (t - 1) - [a, 1.134] \int_t^1 (s - t)(s - 1)\mathrm{d}s$$

$$= (t - 1) - \left[\frac{a}{6}, 0.189\right] (t - 1)^3\,,$$

and we will have $Q_2(X_0)(t) \subseteq X_0(t)$ for all t in $[0, 1]$ provided that

$$(t - 1)\left\{1 - \left[\frac{a}{6}, 0.189\right](t - 1)^2\right\} \subseteq [a, 1](t - 1) \text{ for all } t \text{ in } [0, 1].$$

The relation above will be satisfied for $a > 0$ such that $1 - 0.189(t - 1)^2 \geqslant a$, t in $[0, 1]$. Hence, we may take $a = 0.811$. Then we will have

$$w_2(t) \subseteq (t - 1)\,\{1 - [0.135, 0.189](t - 1)^2\} \subseteq [0.811, 1](t - 1).$$

Note that $w_2(t)$ is negative while $w_1(t)$ is positive for $0 < t < 1$.

Using the interval polynomial containing w_1 and the differential equation for w_1, we find that

$$w_1''(t) \subseteq -[1, 1.134][0.811, 1]\,t$$

$$w_1''(t) \subseteq -[0.811, 1.134]\,t$$

and $$w_1'(t) = 1 + \int_0^t w_1''(s)\mathrm{d}s$$

so $$w_1'(t) \subseteq 1 - [0.4, 0.567]\,t^2\,.$$

Similarly, we find that

$$w_2''(t) \subseteq -[0.811, 1.134](t - 1)$$

and $$w_2'(t) \subseteq 1 - [0.4, 0.567](t - 1)^2\,.$$

We find, furthermore, that

$$\int_0^t |w_1(s)|\,ds \subseteq \int_0^t [0.811, 1]\, s\,ds \subseteq [0.4, 0.5]\,t^2$$

and $$\int_t^1 |w_2(s)|\,ds \subseteq \int_t^1 [0.811, 1](1-s)\,ds \subseteq [0.4, 0.5](1-t)^2 .$$

We find that

$$\frac{d^2}{dt^2}\int_0^1 |G(t, s)|\,ds \text{ is contained in the interval}$$

$$\begin{aligned} B(t) = \frac{1}{[0.811, 0.865]} &\{[0.811, 1.134](1-t)\,[0.4, 0.5]\,t^2 \\ &+ [0.811, 1.134]\,t\,[0.4, 0.5](1-t)^2 \\ &+ (1 - [0.4, 0.567](1-t)^2)\,[0.811, 1]\,t \\ &- (1 - [0.4, 0.567]\,t^2)\,[0.811, 1](1-t)\} . \end{aligned}$$

At a given value of t in $[0, 1]$ the largest number contained in $B(t)$ is the right hand end point of the interval $B(t)$, which is given explicitly by the expression

$$\begin{aligned} r(t) &= \frac{1}{0.811}\{0.567\,(1-t)t + 1 - 0.4(1-t)^2\,t \\ &\quad - (1 - 0.567t^2)\,0.811\,(1-t)\} \\ &= \frac{1}{0.811}\{1 + (1-t)\,(-0.811 + 0.167t + 0.859637t^2)\} \end{aligned}$$

Evaluating the last expression above using interval arithmetic, replacing t by the interval $[0, 1]$, we find that

$$\max_{t\in[0,1]} r(t) \leqslant \frac{1.215637}{0.811} < 1.5 .$$

From this result we conclude that $\|[P'(x_0)]^{-1}\| < 1.5$. Thus, we can take $B_0 = 1.5$.

In this example, then, we have $\|x_1 - x_0\| \leqslant B_0\|P(x_0)\| < (1.5)\,(0.134) = 0.201$, and we can take $\eta_0 = 0.201, B_0 = 1.5$ in the Kantorovich conditions for the operator $P(u) = u'' - e^{-u}$ on $C_0^{(2)}[0, 1]$ with $x_0(t) = \frac{1}{2}t(t-1)$.

To find, K, we observe that

$$[P'(x)y](t) = y''(t) + e^{-x(t)}y(t)$$

and so $$[P''(x)t]u(t) = -e^{-x(t)}y(t)u(t).$$

Thus, we have $[P''(x)y]\,u(t) \;=\; -\,\mathrm{e}^{-x(t)}y(t)u(t)$.

and $$\|P''(x)\| \;=\; \sup_{\|y\|=1} \|P''(x)y\|\,,$$

where $$\|y\| = \max_{t\in[0,1]} |y''(t)|\,,$$

and $$\|P''(x)y\| \;=\; \sup_{\|u\|=1} \|[P''(x)y]\,u\|\,,$$

where $$\|u\| = \max_{t\in[0,1]} |u''(t)|\,,$$

and $$\|[P''(x)y]\,u\| = \max_{t\in[0,1]} |\mathrm{e}^{-x(t)}y(t)u(t)|\,.$$

Recall from Exercise 100, c) 'Hint' that $\|u\|_{C[0,1]} \leqslant \|u\|_{C_0^{(2)}[0,1]}$.
Thus, in this example,

$$\|P''(x)\| \;=\; \max_{t\in[0,1]} |\mathrm{e}^{-x(t)}| \leqslant \mathrm{e}^{\delta}\,\mathrm{e}^{0.125}$$

for $$\|x - x_0\|_{C[0,1]} \leqslant \|x - x_0\|_{C_0^{(2)}[0,1]} \leqslant \delta.$$

Putting $K = 1.134\mathrm{e}^{\delta} > \mathrm{e}^{0.125}\,\mathrm{e}^{\delta}$, we can satisfy the Kantorovich conditions with $\eta_0 = 0.201, B_0 = 1.5, K = 1.134\mathrm{e}^{\delta}$ provided that there is a $\delta > 0$ such that

$$\delta \geqslant \frac{1 - \sqrt{1 - 2h_0}}{h_0}\;(0.201)$$

and $$h_0 \;=\; (1.5)\,(0.201)\,(1.134)\,\mathrm{e}^{\delta}$$

$$= 0.347\ldots\,\mathrm{e}^{\delta} \leqslant \frac{1}{2}.$$

With $h_0 = 0.375 \leqslant \dfrac{1}{2}$, for instance, we can shoose $\delta = 0.268$, and we will have

$$(0.347\ldots)\mathrm{e}^{0.268} \;=\; (0.347\ldots)\,(1.31\ldots) = 0.45\ldots < 0.5.$$

Therefore, the Kantorovich theorem guarantees the convergence of Newton's method to a solution x of $x'' - \mathrm{e}^{-x} = 0, x(0) = x(1) = 0$ from the starting point $x_0(t) \;=\; \dfrac{1}{2}t(t\;-\;1)$. Furthermore, the solution x will be in $N_{\delta}(x_0)$ with δ = 0.268, thus for every t in $[0,\,1]$ we will have

$$|x(t) - x_0(t)| \leqslant \|x - x_0\|_{C[0,1]} \leqslant \|x - x_0\|_{C_0^{(2)}[0,1]} \leqslant 0.268,$$

and it follows that

$$|x(t) - x_1(t)| \leqslant \|x - x_1\| \leqslant \frac{B_0 K}{2} \|x - x_0\|^2$$

$$\leqslant \frac{1.485 \ldots (1.5)}{2} (0.268)^2$$

$$< 0.082, \text{ for all } t \text{ in } [0, 1].$$

For x_2, we will have $\|x - x_2\| < 1.2(0.082)^2 < 0.0081$, and for x_3, $\|x - x_3\| < 1.2(0.0081)^2 < 0.000079$, and for x_4, $\|x - x_4\| < 1.2(0.000079)^2 < 0.0000000075$. The Newton sequence is rapidly convergent here. Of course, each iteration requires the solution (or at least approximate solution) of a linear equation.

Exercise 102 Study the application of Newton's method to the solution of the nonlinear two-point boundary value problem

$$x''(t) - 6(x(t) + 1 + 2.252t)^2 - 5t = 0$$

$$x(0) = x(0.5) = 0 .$$

Consider the operator $P: C_0^{(2)}[0, 0.5] \to C[0, 0.5]$ defined by $P(u)(t) = u''(t) - 6(u(t) + 1 + 2.252t)^2 - 5t$ where $C_0^{(2)}[0, 0.5]$ is the Banach space of twice continuously differentiable functions on $[0, 0.5]$ which vanish at 0 and at 0.5 with norm

$$\|u\| = \max_{t \in [0, 0.5]} |u''(t)| .$$

Find $P'(y)$ and $P''(y)$ and seek an element x_0 and a neighborhood $N_\delta(x_0)$ for which the Kantorovich conditions are satisfied.

Newton's method for nonlinear operator equations in Banach spaces requires the solution (or approximate solution) of a linear operator equation at each iteration. Methods for the approximate solution of linear operator equations have been discussed in Chapters 11–13. Additional methods can be found in Chapters 14 and 15. There exist other methods as well which are not discussed in this work such as factorization (or 'splitting') methods for second order differential equations, shooting methods for two-point boundary value problems, transform methods, and others. Methods based on complex analysis and eigenvalue problems for operators have also been omitted from this work (except for the spectral theory of completely continuous self-adjoint operators in a separable Hilbert space, which does not require complex analysis).

An *interval version* of Newton's method for the class of two point boundary value problems of the form $y''(t) = f(t, y(t))$, $y(0) = y(1) = 0$ has been developed by T. Talbot: 'Guaranteed error bounds for computer solutions of nonlinear two-point boundary value problems', *Mathematics Research Center Technical Summary Report 875*, 1968, Mathematics Research Center, University of Wisconsin–Madison.

A much more general and precise method which combines interval analysis with the theory of *bounded semi-norms* has been developed by Y.-D. Lee in his PhD thesis at the University of Wissonsin–Madison: 'Guaranteed component-wise, point-wise error bounds for approximate solutions of nonlinear two-point boundary value problems using an improved Kantorovich-like theorem', 1980.

A method which can be even more rapidly convergent than Newton's method and which uses second Fréchet derivatives is 'Chebyshev's method'. See: R. E. Moore and Shen Zuhe, 'An interval version of Chebyshev's method for nonlinear operator equations', *J. Nonlinear Analysis* 7, no. 1, 1983, pp. 21–34. This method as well as others in this work, can be applied to certain operator equations involving nonlinear partial differential equations.

CHAPTER 18

Variants of Newton's method

In this chapter, a general theorem is proved, which includes as special cases: a theorem of Ostrowski, Newton's method, the simplified Newton method, and the successive over-relaxation Newton method. Numerical examples illustrate these variants of Newton's method.

A number of references to further works on such methods can be found at the end of the chapter.

Suppose that B_1 and B_2 are Banach spaces and that $P: N_\delta(x) \subseteq B_1 \to B_2$ is an operator for which $P(x) = 0$ and $P'(x)$ exists. Consider iterative methods for approximating x, $x_{k+1} = F(x_k)$, $(k = 0, 1, 2, \ldots)$ where F can be expressed as $F(y) = y - A(y)\, P(y)$; where, for every y in $N_\delta(x)$, $A(y)$ is a bounded linear operator from B_2 into B_1 and $\|A(y)\|$ is uniformly bounded, say $\|A(y)\| \leqslant \alpha$ for all y in $N_\delta(x)$.

We have the following.

Theorem If $\|A(y)P'(x) - I\| \leqslant \theta < 1$ for every y in $N_\delta(x)$, then there is a $\delta' > 0$ such that the iterative method above converges to x from every x_0 in $N_{\delta'}(x)$.

Proof Since $P'(x)$ exists and $P(x) = 0$ we can write

$$\begin{aligned} P(y) &= P(x) + P'(x)\,(y-x) + G(x, y) \\ &= P'(x)\,(y-x) + G(x, y) \end{aligned}$$

where, for every $\epsilon > 0$, there is a $\delta(\epsilon) > 0$ such that $\|G(x, y)\| \leqslant \epsilon\|x - y\|$ whenever y is in $N_{\delta(\epsilon)}(x)$. Then $F(y) - x = y - x - A(y)P'(x)\,(y - x) - A(y)\, G(x, y)$, and

$$\begin{aligned} \|F(y) - x\| &\leqslant \|I - A(y)P'(x)\|\,\|y - x\| + \alpha\,\epsilon\,\|y - x\| \\ &\leqslant (\theta + \alpha\epsilon)\,\|y - x\|\,. \end{aligned}$$

Choose $\quad \epsilon < \dfrac{1-\theta}{\alpha}$, then $\|F(y) - x\| \leqslant c\|y - x\|$ for every y in $S_0 =$

$N_{\delta(\epsilon)}(x)$ and $c = \theta + \alpha\epsilon < 1$. Thus, F maps the set S_0 with radius $\sigma_0 = \delta(\epsilon)$ into the set $S_1 = F(S_0) = \{Z = F(y) : y \in S_0\} \subseteq S_0$ with radius $\sigma_1 = \sup_{z \in S_1} \|z - x\| \leqslant c\ \sigma_0$. Similarly, the sets $S_k = F(S_{k-1}) \subseteq S_{k-1}$, $k = 1, 2, \ldots$, have radii $\sigma_k \leqslant c^k\ \sigma_0$. Therefore the iterative method produces a sequence of points $\{x_k\}$ in S_0 converging to x from every x_0 in S_0, and the theorem is proved.

Corollary Under the conditions of the theorem, we also have

$$\|x_k - x\| \leqslant \frac{c^k}{1-c} \|x_0 - x_1\|$$

where $c = \theta + \alpha\epsilon < 1$.

The theorem covers several cases of special interest.

Example 1 (cf: Ostrowski theorem; Ortega and Rheinboldt, p. 300)†
For $A(y) \equiv I$ the iterative method has the form $x_{k+1} = F(x_k) = x_k - P(x_k)$. In this case, $F'(x)$ exists whenever $P'(x)$ exists, and the theorem states that if $F(x) = x$ and $\|F'(x)\| < 1$, then the iteration converges to x from every x_0 in some neighborhood of x.

Example 2
For $A(y) = [P'(y)]^{-1}$, assuming that $[P'(y)]^{-1}$ exists and that $\|[P'(y)]^{-1}\| \leqslant \alpha$ for y in $N_\delta(x)$, we have *Newton's method*

$$x_{k+1} = F(x_k) = x_k - [P'(x_k)]^{-1} P(x_k).$$

In this case, the theorem asserts that if $\|[P'(y)]^{-1} P'(x) - I\| \leqslant \theta < 1$ when y is in $N_\delta(x)$, then there is a $\delta' > 0$ such that Newton's method converges from any x_0 in $N_{\delta'}(x)$.

Example 3
For $A(y) \equiv [P'(x_0)]^{-1}$, assuming that $[P'(x_0)]^{-1}$ exists, we have the *simplified Newton's method*

$$x_{k+1} = F(x_k) = x_k - [P'(x_0)]^{-1} P(x_k).$$

If $\|[P'(x_0)]^{-1} P'(x) - I\| \leqslant \theta < 1$ then for some $\delta' > 0$ the iteration converges to x from any x_0 in $N_{\delta'}(x)$. For an analysis and illustration of this method see Rall (pp. 198–203) (see refs at end of this chapter).

Example 4
A practical method for $B_1 = B_2 = E^n$, referred to by Ortega and Rheinboldt as

† See references at the end of this chapter.

the *SOR* (*successive over-relaxation*) – *Newton Method* and by Greenspan as the *generalized Newton Method*, is the following.

Suppose that the $n \times n$ matrix $P'(y)$ has *nonzero diagonal elements* $\dfrac{\partial P_i}{\partial y_i}(y)$ for y in $N_\delta(x)$. Then take $A(y)$ to be the diagonal matrix whose nonzero elements are

$$[A(y)]_{ii} = w \left[\frac{\partial P_i}{\partial y_i}(y)\right]^{-1}.$$

The iterative method that results in component form, is

$$x_i^{(k+1)} = x_i^{(k)} - w \left[\frac{\partial P_i}{\partial y_i}(x^{(k)})\right]^{-1} P_i(x^{(k)}), (i = 1, 2, \ldots, n)$$

A *Gauss–Seidel modification* of the method is employed most commonly (see Greenspan) in which $x_i^{(k+1)}$ is used instead of $x_j^{(k)}$ on the right hand side of the iteration formula for $j = 1, 2, \ldots, i-1$. This has the effect of not requiring separate computer storage for the vectors $x^{(k)}$ and $x^{(k+1)}$. The components $x_i^{(k+1)}$ are stored on top of $x_i^{(k)}$ as soon as they are computed.

The SOR–Newton method is simpler notationally for analysis than its Gauss–Seidel modification and has similar convergence properties. See Ortega and Rheinboldt for a thorough analysis of the Gauss–Seidel modification.

Let $M(y)$ be the matrix whose elements are

$$[M(y)]_{ii} = \left[\frac{\partial P_i}{\partial y_i}(y)\right]^{-1} \frac{\partial P_i}{\partial y_i}(x) - 1$$

$$[M(y)]_{ij} = \left[\frac{\partial P_i}{\partial y_i}(y)\right]^{-1} \frac{\partial P_i}{\partial y_j}(x), \ i \neq j$$

$i, j = 1, 2, \ldots, n$. Then $A(y)P'(x) - I = wM(y) + (w-1)I$ and $\|A(y)P'(x) - I\| \leqslant |w|\,\|M(y)\| + |w-1|$.

From the above inequality and the theorem of this chapter, it follows that the SOR–Newton method converges

1) for any $0 < w \leqslant 1$ if $\|M(y\|\| < 1$ for all y in $N_\delta(x)$

2) for any $1 \leqslant w < 2$ if $\|M(y\|\| < \dfrac{2}{w} - 1$ for all y in $N_\delta(x)$ from every x_0 in some neighborhood of x.

If $\left[\dfrac{\partial P_i}{\partial y_i}(y)\right]^{-1}$ is continuous in y for y in some neighborhood of x, then

$[M(y)]_{ii}$ is near zero and, for the maximum row sum norm; we have, in fact, for any $\epsilon > 0$

$$\|M(y)\| \leqslant \max_{i=1,2,\ldots,n} \sum_{\substack{j=1 \\ j\neq i}}^{n} \left| \left[\frac{\partial P_i}{\partial y_i}(x)\right]^{-1} \frac{\partial P_i}{\partial y_j}(x) \right| + \epsilon$$

whenever y is in some $\delta(\epsilon)$ neighborhood of x. Thus if

$$\sum_{j\neq i} \left| \frac{\partial P_i}{\partial y_j}(x) \right| < \left| \frac{\partial P_i}{\partial y_i}(x) \right| \quad (i = 1, 2, \ldots, n)$$

then $\|M(y)\| < 1$ for all y in some neighborhood of x. The condition above is that of *diagonal dominance* of the matrix $P'(x)$. For example, if we discretize the operator $P(u) = u'' - e^{+u}$ as $P_i(y) = y_{i+1} - 2y_i + y_{i-1} - \Delta t^2 e^{+y_i}$ where $\Delta t = 1/N$, with $y = (y_1, y_2, \ldots, y_{N-1})$ and apply the boundary conditions $y_0 = y(0) = y_N = y(1) = 0$, then $P'(x)$ is a tridiagonal matrix with

$$\frac{\partial P_i}{\partial y_i}(x) = 0 \text{ for } |j - i| > 1,$$

and the condition becomes $2 < 2 + \Delta t^2 \exp(x_i)$ $(i = 1, 2, \ldots, N)$.

Thus the SOR–Newton iterative method will converge for $P(y)$ given as above for any $0 < w \leqslant 1$. For the operator $P(u) = u'' - e^{-u}$, however, the same argument does not work, and we cannot conclude from the analysis given here that the SOR-Newton method will converge to a zero of the discretized version of the boundary value problem $u'' - e^{-u} = 0, u(0) = u(1) = 0$.

Numerical examples

The two-point boundary value problem $y'' = 6y^2 + 5t$ with $y(0) = 1.0, y(0.5) = 2.126$ was discretized with $\Delta t = 1/2N$ and the differential equation replaced by $P(y) = y_{i+1} - 2y_i + y_{i-1} - \Delta t^2 (6y_i^2 + 5t_i) = 0, i = 1, 2, \ldots, N-1$, for $y = (y_1, y_2, \ldots, y_{N-1})$ with $t_i = i\Delta t$ and y_i approximating $y(t_i)$ for $i = 1, 2, \ldots, N-1$. The boundary conditions were imposed as $y_0 = 1.0$ and $y_N = 2.126$. The Gauss–Seidel modification of the SOR-Newton method was used to solve the discretized equations. For this example we have

$$\frac{\partial P_i(y)}{\partial y_i} = -2 - 12\Delta t^2 y;$$

so that the iterations take the form

$$y_i^{(k+1)} = y_i^{(k)} - w \left\{ \frac{y_{i+1}^{(k)} - 2y_i^{(k)} + y_{i-1}^{(k+1)} - \Delta t^2 \{6(y_i^{(k)})^2 + 5t_i\}}{-2 - 12\Delta t^2 y_i^{(k)}} \right\}$$

$i = 1, 2, \ldots, N-1$. Using $w = 1$ and the initial approximation $y_i^{(0)} = 1 + 2.252t_i$, $i = 1, 2, \ldots, N-1$ (a straight line passing through the boundary values), the iterations were carried out until

$$\max_{i=1,2,\ldots,N-1} |y_i^{(k+1)} - y_i^{(k)}| = \max_{i=1,2,\ldots,N+1} \frac{|P_i(y)|}{\left|\dfrac{\partial P_i}{\partial y_i}(y)\right|} \leqslant 10^{-4}$$

was satisfied. The following table shows the number of iterations that were required to satisfy the stopping criterion as a function of N:

N	10	20	30	40	50
# iterations	47	143	265	403	547

The program was written in FORTRAN IV for the UNIVAC 1110 computer at the University of Wisconsin. The entire computation summarized by the table required 3.6 seconds of machine time. For comparison, the equations were then solved using Newton's method and taking advantage of the tridiagonal form of the resulting linear systems. For the same initial approximation and the comparable stopping criterion

$$\max_{i=1,2,\ldots,N-1} |y_i^{(k+1)} - y_i^{(k)}| \leqslant 10^{-4}$$

the Newton method stopped in only a few iterations in all cases $N = 10, 20, 30$ $40, 50$. The total computing time for Newtons method was 1.9 seconds.

While no one is going to get very excited about saving 1.7 seconds of computing time it is, at least, interesting to note that Newton's method *was faster* than the SOR-Newton method in this example in spite of having to solve a linear system at each iteration. Of course, since only one value of w was tried, it may be that the comparison was not completely fair. Perhaps another value of w would yield faster convergence for the SOR-Newton method in this example.

Exercise 103 Using the information given by the stopping criterion used and the table following that relation, *estimate* the accuracy obtained with $w = 1$ for $N = 10$. Try the SOR-Newton method with other values of w.

For extensive discussions on the determination of an *optimal* value for the relaxation factor w for *linear* systems of equations see Young (1971) and Forsythe and Wasow (1960). For computational results on partial differential equations using the SOR-Newton method or its Gauss–Seidel modification see Greenspan (1968, 1974).

Note that in the SOR-Newton method each component $x_i^{(k+1)}$ is found explicitly. The coefficient $\frac{\partial P_i}{\partial y_i}(x^{(k)})$ is a number, and only an ordinary division need be performed. No linear *system* need be solved at each iteration as in the Newton method. For very large systems of nonlinear equations such as arise in finite differece methods for nonlinear partial differential equations, for example Greenspan (1968), there may sometimes be a computational advantage in using the SOR-Newton method or its Gauss–Seidel modification instead of Newton's method. On the other hand, the matrices occurring in the successive linear systems to be solved in such an application will be *sparse* (mostly zero elements) so that Gaussian elimination – just as in the tridiagonal case – can be arranged in such a way as to take advantage of the large number of zero coefficients to vastly reduce the amount of computation required.

References

D. M. Young, *Iterative solution of large linear systems*, Academic Press, New York, 1971.

E. Isaacson and H. B. Keller, *Analysis of numerical methods*, John Wiley & Sons, Inc., New York 1966.

J. M. Ortega and W. C. Rheinboldt, *Iterative solution of nonlinear equations in several variables*, Academic Press, New York, 1970.

G. E. Forsythe and W. R. Wasow, *Finite-difference methods for partial differential equations*, John Wiley & Sons, Inc., New York, 1960.

D. Greenspan,
1) *Lectures on the numerical solution of linear, singular and nonlinear differential equations*, Prentice-Hall, Inc. Engelewood Cliffs, N. J., 1968
2) *Discrete numerical methods in physics and engineering*, Academic Press, New York, 1974.

L. B. Rall, *Computational solution of nonlinear operator equations*, John Wiley & Sons, New York, 1969; reprinted by Krieger, Huntington, N.Y., 1979.

P. M. Anselone (Ed.), *Nonlinear integral equations*, University of Wisconsin Press, 1964 (especially: R. H. Moore, 'Newton's method and variations,' pp. 65–98).

L. B. Rall (Ed.) *Nonlinear functional analysis and applications*, Academic Press, 1971 (especially: J. E. Dennis, Jr., 'Toward a unified convergence theory for Newton-like methods,' pp. 425–472).

CHAPTER 19

Homotopy and continuation methods

For iterative methods for solving nonlinear operator problems, the difficult and important problem remains of finding an initial guess that is close enough to a solution so that the iterative method converges to the solution. Sometimes this can be done by using additional information about the operator equation derived from close examination of the particular operator equation involved, or from a crude approximation to a solution found by other means.

An elegant approach to the problem of finding a sufficiently close initial approximation for a given iterative method to converge is afforded by 'continuation' methods based on the concept of a homotopy. In this chapter, we give an introductory treatment of this approach. The basic idea is this: we write an operator equation that we can solve which somehow resembles the one we want to solve. We then find a continuous transformation (a homotopy) depending on a parameter λ such that, for $\lambda = 0$, we have the problem we *can* solve, and for $\lambda = 1$ we have the problem we *want* to solve. Thus, we have a continuous transformation of a solvable problem into the problem of interest. We then subdivide the interval $[0, 1]$ into a finite number of sub-intervals, and take the solution (or an approximate solution) at λ_i as an initial guess for the solution of the problem at λ_{i+1} in some iterative method. We hope, in this way, to find a suitable initial guess for the original problem, which corresponds to $\lambda = 1$. In practice, the approach is most successful if we take a problem corresponding to $\lambda = 0$ which closely resembles the problem we want to solve, but has a known solution.

References to research papers on the continuation approach are given.

An alternative approach, in the case of finite dimensional nonlinear operator equations, is given in Chapter 6 of *Methods and applications of interval analysis* (R. E. Moore, SIAM, Philadelphia, 1979).

A family of continuous mappings $h_\lambda : X \to Y, \lambda \in [0, 1]$ is called a *homotopy* if the function

$H : X \times [0, 1] \to Y$ defined by

$$H(x, \lambda) = h_\lambda(x), x \in X, \lambda \in [0, 1]$$

is continuous (with the product space topology on $X \times [0, 1]$). (See S. T. Hu, *Homotopy theory*, Academic Press, 1959). Here X and Y can be any two topological spaces (for instance Banach spaces). The maps h_0 and h_1 are called, respectively, the *initial map* and the *terminal map* of the homotopy h_λ. Two maps, $f: X \to Y$ and $g: X \to Y$, are said to be *homotopic* if there exists a homotopy h_λ such that $h_0 = f$ and $h_1 = g$. Then f can be changed continuously into g.

In this chapter we consider the use of homotopies in *successive perturbation methods*, often called *continuation methods*.

By an *open set* S in a Banach space B we mean a subset S of B such that every point y in S has a $\delta(y) > 0$ neighborhood contained in S. Thus S is open if y in S implies that, for some $\delta(y) > 0, N_{\delta(y)}(y) = \{y : z \text{ in } B, \|z - y\| \leqslant \delta(y)\} \subseteq S$. Note that B is open (in itself).

Suppose we are interested in finding a zero of a continuous mapping $P: S \subseteq B_1 \to B_2$ where S is an open set in B_1. Let x_0 be any element of S and consider the homotopy $h_\lambda(y) = H(y, \lambda) = P(y) + (\lambda - 1)P(x_0)$ with initial map $h_0(y) = P(y) - P(x_0)$ and terminal map $h_1(y) = P(y)$. Clearly, h_0 has a zero at $y = x_0$.

Under various conditions, $H(y, \lambda)$ will have a zero for each λ in $[0, 1]$, and the zero x_0 of $H(y, 0)$ may lead to an approximation to a zero of $H(y, 1) = P(y)$ if we can approximately follow a *curve of zeros* $x(\lambda)$ satisfying $H(x(\lambda), \lambda) \equiv 0, \lambda$ in $[0, 1]$, with $x(0) = x_0$.

We can write $H(y, \lambda) = Q(y) + \lambda P(x_0)$, where $Q(y) = P(y) - P(x_0)$.

Suppose that, for every y in S, $P'(y)$ exists and $A(y)$ is a bounded linear operator from B_2 into B_1, and that $\|A(y)\| \leqslant \alpha$ for all y in S. If there is a continuous curve of zeros $x(\lambda)$ in the open set S, then we have the following *discrete continuation* method.

Theorem If, for each λ in $[0, 1]$, there is a $\delta(\lambda) > 0$ such that $\|A(y)P'(x(\lambda)) - I\| \leqslant \theta < 1$ for every y in $N_{\delta(\lambda)}(x(\lambda))$ (with θ independent of λ and y), then for any $\epsilon > 0$ there are integers $M, N_1, N_2, \ldots, N_M$ and numbers $0 < \lambda_1 < \lambda_2 < \ldots < \lambda_M = 1$ such that $\|x(1) - x_{M,N_M}\| \leqslant \epsilon$ where

$$x_{1,0} = x_0$$

$$x_{j,k+1} = F_j(x_{j,k}), \quad k = 0, 1, \ldots, N_{j-1}$$

$$x_{j+1,0} = x_{j,N_j}, \quad j = 1, 2, \ldots, M - 1$$

with F_j defined by $F_j(y) = y - A(y)H(y, \lambda_j), (j = 1, 2, \ldots, M)$.

Proof From the theorem of Chapter 18, for any λ_j in $[0, 1]$ there is a $\delta'(\lambda_j) > 0$ such that $x_{j,k+1} = F_j(x_{j,k}), k = 0, 1, \ldots$ converges to $x(\lambda_j)$ from any $x_{j,0}$ in $\{y : \|y - x(\lambda_j)\| \leqslant \delta'(\lambda_j)\}$. From the assumed continuity of $x(\lambda)$ on the compact set [0, 1] it follows that there is a positive lower bound $0 < \delta \leqslant \delta'(\lambda_j)$ for λ_j in [0, 1]. From the uniform continuity of $x(\lambda)$ on [0, 1] we can choose $\Delta\lambda$ small enough so that $\|x(\lambda_{j+1}) - x(\lambda_j)\| < \delta$ for $0 < \lambda_{j+1} - \lambda_j \leqslant \Delta\lambda$. Then for some integer M there are $0 < \lambda_1 < \lambda_2 < \ldots < \lambda_M = 1$ with $\lambda_{j+1} - \lambda_j \leqslant \Delta\lambda$ such that

$$x(\lambda_j) \in S_{\lambda_{j+1}} = \{y : \|y - x(\lambda_{j+1})\| \leqslant \delta\}.$$

Thus $x_{j+1,k+1} = F_{j+1}(x_{j+1,k})$ converges to $x(\lambda_{j+1})$ from any $x_{j+1,0}$ that is close enough to $x(\lambda_j)$ to be in $S_{\lambda_{j+1}}$. This implies the existence of the integers $N_1, N_2, \ldots, N_M$ in the conclusion of the theorem, and the theorem is proved.

With further assumptions for specific algorithms we can obtain more precise information. Let us consider, for example, in more detail such an argument as applied to Newton's method.

Suppose that $\|P^{(2)}(y)\| \leqslant K$ and $\|[P'(y)]^{-1}\| \leqslant B$ for all y in S. This time, *without assuming* the existence of zeros of $H(y, \lambda)$ in advance, let us seek sufficient conditions to guarantee that a zero exists and that x_0 is a safe starting point for the convergence of the Newton iterates to a zero $x(\lambda_1)$ of $H(y, \lambda_1)$ for some $\lambda_1 > 0$. In fact, let us seek conditions such that $x(\lambda)$ is a safe starting point for convergence to a zero $x(\lambda + \Delta\lambda)$ of $H(y, \lambda + \Delta\lambda)$ if $H(x(\lambda), \lambda) = 0$.

In order to apply the Kantorovich conditions (see Chapter 17) here, with $H(y, \lambda + \Delta\lambda) = H(y, \lambda) + \Delta\lambda P(x_0)$, and $H(y, \lambda) = P(y) + (\lambda - 1)P(x_0)$, we can take $\eta \geqslant B|\Delta\lambda|\ \|P(x_0)\| \geqslant \|x_1 - x(\lambda)\|$, where $x_1 = x(\lambda) - [H'(x(\lambda), \lambda + \Delta\lambda)]^{-1} H(x(\lambda), \lambda + \Delta\lambda)$, since $H'(y, \lambda) = P'(y)$, and $H(x(\lambda), \lambda + \Delta\lambda) = \Delta\lambda P(x_0)$.

Now, for $h = B\eta K \leqslant \frac{1}{2}$, we can require that $h_0 = B^2\ |\Delta\lambda|\|P(x_0)\|\,K < \frac{1}{2}$.

Finally, we require that $N_\delta(x(\lambda)) \subseteq S$ where

$$\delta \geqslant \frac{1 - \sqrt{1 - 2h}}{BK}.$$

For small enough $\Delta\lambda > 0$ we can satisfy the above inequality with

$$\delta \geqslant \delta_0 = \frac{1 - (1 - 2B^2\,|\Delta\lambda|\,\|P(x_0)\|\,K)^{1/2}}{BK} > 0$$

Thus, if $N_\delta(x_0) \subseteq S$ for δ satisfying the above inequality, and if $h_0 = B^2\ |\Delta\lambda|\ \|P(x_0)\|\ K < 1/2$ holds, then the Kantorovich theorem applies and $H(y, \Delta\lambda)$ has a zero in $N_\delta(x_0)$ to which the Newton iterates converge. Call $\lambda_1 = \Delta\lambda$ and denote by $x(\lambda_1)$ the zero of $H(y, \lambda_1)$ in $N_\delta(x_0)$. We have $H(x(\lambda_1), \lambda_1) = 0$.

If again $N_\delta(x(\lambda_1)) \subseteq S$ for some δ as above, then again the Kantorovich theorem applies, and Newton's method will converge to a zero $x(\lambda_2)$ of $H(y, \lambda_2) = 0$ in $N_\delta(x(\lambda_1))$ where $\lambda_2 = \lambda_1 + \Delta\lambda$. Furthermore, since $\delta > 0$, the Newton iterates will also converge to $x(\lambda_2)$ from an *approximation* to $x(\lambda_1)$ (in fact, from any point in $N_\delta(x(\lambda_1))$. Finally, we have the following result.

Theorem If there is a $\Delta\lambda > 0$ such that $h_0 < 1/2$ and $N_{\delta_j}(x(\lambda_j)) \subseteq S$ for some δ_j satisfying $\delta_j \geqslant \delta_0$, $\lambda_{j+1} = \lambda_j + \Delta\lambda$, then the conclusions of the previous theorem hold, putting $A(y) = [P'(y)]^{-1}$; given an $\epsilon > 0$, a finite series of Newton iterations will produce an approximation x_{M,N_M} to a zero $x(1)$ of P in S such that $\|x(1) - x_{M,N_M}\| \leqslant \epsilon$.

Example

Let B_1 and B_2 be the real line and consider the mapping $P(y) = \ln(1 + y)$ defined and continuous on the open set $S = (-b, a)$ for any $0 < b < 1, a > 0$. We have

$$[P'(y)]^{-1} = 1 + y$$

$$P^{(2)}(y) = \frac{-1}{(1+y)^2},$$

so we can take $K = \dfrac{1}{(1-b)^2} \geqslant |P^{(2)}(y)|,\ y \in S$

$$B = 1 + a \geqslant |[P'(y)]^{-1}|,\ y \in S\,.$$

The exact region of safe starting points for Newton's method to converge to the zero of P at $y = 0$ is easily found to be the open interval $(-1, e - 1)$. The Kantorovich conditions will be satisfied in some subinterval.

Using the theorem above with $H(y, \lambda) = \ln(1 + y) + (\lambda - 1)\ln(1 + x_0)$ for any $S = (-b, a)$ with $0 < b < 1, a > 0$ and any x_0 in S, we can satisfy $h_0 < 1/2$ and $\delta \geqslant \delta_0$ with

$$\Delta\lambda < \frac{(1-b)^2}{2\,|\ln(1+x_0)|\,(1+a)^2}$$

and $$-b < x(\lambda) - \delta < x(\lambda) + \delta < a$$

where $$\delta \geqslant \frac{1 - \left(1 - \dfrac{2(1+a)^2}{(1-b)^2}\,\Delta\lambda\,\ln(1+x_0)\right)^{1/2}}{\dfrac{1+a}{(1-b)^2}} > 0.$$

For any $-1 < -b < x_0 < a < \infty$ we can take $\Delta\lambda$ sufficiently small to satisfy the above inequalities as long as $x(\lambda)$ remains in the interval $(-b, a)$.

We can check directly in this simple example that $H(x(\lambda), \lambda) = 0$ for $\ln(1 + x(\lambda)) + (\lambda - 1)\ln(1 + x_0) = 0$, thus $x(\lambda) = (1 + x_0)^{1-\lambda} - 1$. We can check that $x(0) = x_0$, $x(1) = 0$, and $x(\lambda)$ lies between x_0 and 0 in this example.

Thus, the discrete continuation method, using Newton's method ($A(y) = [P'(y)]^{-1}$) at each step, in this example, can be made to converge from any $x_0 > -1$ to the zero of P. Whereas *the ordinary sequence of Newton iterates converges from* x_0 *to a zero of* P *only if* $-1 < x_0 < e - 1$.

DAVIDENKO'S METHOD:
Another approach to the determination of a curve of zeros leading from some x_0 to a desired $x(1)$ is as follows.

Differentiating $H(x(\lambda), \lambda) \equiv 0$ with respect to λ, using $H(y, \lambda) = P(y) + (\lambda - 1)P(x_0)$, we obtain $P'(x(\lambda))x'(\lambda) + P(x_0) = 0$ (assuming that a *differentiable* curve of zeros exists, for the moment). We can rewrite this as the differential equation $x'(\lambda) = -[P'(x(\lambda))]^{-1} P(x_0)$ which, together with $x(0) = x_0$, can be viewed as an initial value problem. If we can find an approximate solution at $\lambda = 1$ then this can be taken as an approximation to a zero of P, since

$$P(x(1)) = P(x(0)) + \int_0^1 P'(x(\lambda))x'(\lambda)\,d\lambda .$$

Davidenko's method (see L. B. Rall, 'Davidenko's method for the solution of nonlinear operator equations,' *MRC Technical Summary Report #968* Mathematics Research Center, University of Wisconsin–Madison, October 1968) consists of finding an approximate solution to the initial value problem at $\lambda = 1$ by using numerical integration techniques (Runge–Kutta, etc.). For instance, if we apply Euler's method we obtain the algorithm

$$x(\lambda + \Delta\lambda) = x(\lambda) - \Delta\lambda[P'(x(\lambda))]^{-1} P(x_0)$$
$$x(0) = x_0 .$$

If we put $\Delta\lambda = 1/N$ and denote $x_k = x(k\Delta\lambda)$, then $x_{k+1} = x_k - \Delta\lambda[P'(x_k)]^{-1} P(x_0)$, $k = 0, 1, 2, \ldots, N-1$.

For $N = 1$, the algorithm above is just Newton's Method! In any case, we can take the result x_N after N steps as an approximate solution to $P(y) = 0$ obtained starting from x_0.

The above cited report by L. B. Rall contains existence and convergence theorems, computational results, illustrations, translations of Russian papers by Davidenko, and numerous references. Ortega and Rheinboldt (*Iterative solution of nonlinear equations in several variables*, Academic Press, New York, 1970) (especially §7.5 and §10.4 on *continuation methods*) give extensive discussion and references to methods and results of the type discussed in this chapter. Two papers on discrete continuation giving theoretical results similar to the theorems of this chapter are:

Ficken, F.A. 'The continuation method for functional equations.' *Comm. Pure Appl. Math.* **4** (1951), 435–456, *Math. Reviews* **13** (1952) 562–563.

Lahaye, E. 'Sur la résolution des systèmes d'équations transcendantes.' *Acad. Roy. Belgique. Bull. Cl. Sci.* (5) **34** (1948), *Math Reviews* **10** (1949), 626.

Further development of this approach is contained (for operators on E^n) in the paper:

Mayer, G. H., 'On solving nonlinear equations with a one-parameter operator imbedding' *SIAM J. Numer. Anal.* **5**, No. 4, Dec. 1968, pp. 739–752.

COMPUTATIONAL ASPECTS

For efficient computational procedures based on continuation methods using the discrete approach one should probably do something like the following.

To find (approximately) a zero $x(1)$ of P using the homotopy $H(y, \lambda) = P(y) + (\lambda - 1) P(x_0)$:

1) Find the largest $(\Delta\lambda)_1$ for which x_0 is a good initial approximation for some iterative method to a zero of $H(y, (\Delta\lambda)_1)$;
2) compute one or a few iterations, obtaining a point x_1;
3) find the largest $(\Delta\lambda)_2$ for which x_1 is a good initial approximation to a zero of $H(y, \lambda_2)$ with $\lambda_2 = (\Delta\lambda)_1 + (\Delta\lambda)_2$;
4) iterate once or twice, obtaining x_2;
5) etc. until $\lambda = 1$ is reached (if possible).

In the general step, then, we would take the result x_i of iteratively approximating a zero of $H(y, \lambda_i)$ using only one or at most a few iterations and use x_i as a starting point to find, in the same way, an approximate zero of $H(y, \lambda_i + (\Delta\lambda)_i)$ for as large as possible $(\Delta\lambda)_i$ until we reach $\lambda = 1$. (See: Rall, MRC #968, *loc. cit.*) Of course, if x_0 is already a safe starting point (even more so if it is a good starting point) for an iterative method for finding a zero of P, then probably nothing will be gained by taking $(\Delta\lambda)_1$ less than 1.

When using the differential equation approach (Davidenko's method) in a step-by-step numerical integration one should probably take as large a step $\Delta\lambda$ as possible to stay somewhere near the curve of zeros until $\lambda = 1$ is reached with an approximation to $x(1)$ which can then be improved if needed by some iterative method such as Newton's method.

In this connection, G. H. Meyer (*loc. cit.*) shows that if $P : E^n \to E^n$ is twice differentiable and satisfies $\|[P'(x)]^{-1} \leqslant \alpha \|x\| + \beta$ for all x in E^n, then for arbitrary x_0 in E^n we have the following.

Suppose
$$x'(t) = -[P'(x(t))]^{-1} P(x_0)$$
$$x(0) = x_0$$

is integrated from $t = 0$ to $t = 1$ with a numerical method of order h^P, say

$x_k \approx x(kh)$ and $x_{k+1} = \Phi(x_k, h)$, and $\|x(1) - x_N\| \leqslant Ch^P$ where C does not depend on $h = 1/N$. Then the iteration

$$x_{k+1} = \Phi(x_k, h), \; k = 1, \ldots, N-1$$

$$x_{k+1} = x_k - [P'(x_k)]^{-1} P(x_k), \; k = N, N+1, \ldots$$

converges to the unique solution of $P(x) = 0$ provided that

$$h = \frac{1}{N} \leqslant \left(\frac{\sqrt{2}-1}{\beta' LC}\right)^{1/p} \quad \text{and } Ch^P < \delta$$

where β' and L are constants such that $\|[P'(x)]^{-1} \leqslant \beta'$ and $\|P^{(2)}(x)\| \leqslant L$ for all x in $N_{r+\delta}(x_0)$ and r is given by

$$r = \begin{cases} \left[\|x_0\| + \dfrac{\beta}{\alpha}\right] \exp(\alpha \|P(x_0)\|) - \|x_0\| - \dfrac{\beta}{\alpha}, & \text{if } \alpha \neq 0 \\ \beta \|P(x_0)\| \text{ if } \alpha = 0 . \end{cases}$$

CHAPTER 20

A hybrid method for a free boundary problem

Real-world problems are often of such complexity that no single mathematical method will suffice. Instead, a variety of methods may be required with *ad hoc* techniques for putting them together. In the end we have what might be called a 'hybrid' method for the problem. We discuss such a problem in this chapter, and design a hybrid method combining: inner products and the Gram–Schmidt process (Chapter 6); separation of variables and eigenfunction expansions (Chapters 12 and 13); and finite difference methods (Chapter 13). A mathematical proof of convergence of the resulting hybrid method is not known to the author, but remains an interesting open research problem.

In this final chapter, we will discuss a problem which arose during the early days of space flight in connection with the 'sloshing' of liquid fuel in cylindrical tanks during the process of restarting the rocket engines while in zero-gravity. It is a problem which is still of some concern.

The method discussed and the results given are taken from the paper 'Inviscid fluid flow in an accelerating cylindrical container', R. E. Moore and L. M. Perko, *J. Fluid Mech.* **22**, part 2, 1965, pp. 305–320, from which Figs. 2–9 are reproduced by kind permission of the editor.

Some background history on the problem might be of interest. The mathematical formulation of the problem involves a partial differential equation (Laplace's equation in cylindrical coordinates) and a free surface condition involving Bernoulli's equation at the interface between liquid and gas. Surface tension must also be taken into account, and the free boundary condition involves a nonlinear partial differential equation to be imposed on a surface whose location is unknown and varying with time. In fact, it is the location and motion of the free surface which we want to find as the solution of the problem.

Previous attempts, by others, to find numerical solutions using only finite difference methods failed; after a year and a half, such attempts were abandoned and a new approach was sought. We will now describe an approach that was successful. It combines separation of variables, eigenfunction expansions, Gram–Schmidt orthonormalization in an inner product space, and a few special tricks that were needed to put together a method that would work, including the introduction of special weight functions in the inner products and finite difference methods used in a special way in a hybrid combination with the eigenfunction expansions.

A complete mathematical theory for the convergence of the method seems out of reach. This is very often the case in real-world problems needing some mathematical analysis. Nevertheless, numerical results showed reasonable behavior and agreement with such experimental measurements as were available.

The problem considered is that of the axially symmetric irrotational flow of an inviscid incompressible fluid with a free surface in a circular cylinder accelerating parallel to its axis.

We will describe a numerical procedure for simulating on a computer the motions of the free surface. Some interesting motions are obtained including the development of breakers near the wall as well as near the axis, splashing, and sustained large amplitude oscillations of the surface.

We may assume that the unit of distance is defined so that the radius of the cylindrical container is 1.0. Let $v(r, z, t)$ be the axially symmetric velocity of a point (r, θ, z) in the fluid at time t. Since the flow is assumed irrotational, there exists a velocity potential $\phi(r, z, t)$. Further, since the flow is assumed incompressible, the velocity potential satisfies Laplace's equation in the interior of the fluid $\phi_{rr} + (1/r)\,\phi_r + \phi_{zz} = 0$, in cylindrical coordinates, for $t \geqslant 0$, $0 < r < 1$, $0 \leqslant \theta < 2\pi$, $0 < z < f(r, t)$ where $z = f(r, t)$ is the (unknown) equation of the free surface.

Since the fluid does not move through the walls of the container, we assume that the normal derivatives are zero on the fixed boundaries; that is, for all $t \geqslant 0$,

$$\begin{aligned}
\phi_r(0, z, t) &= 0 \text{ for } 0 \leqslant z \leqslant f(0, t), \text{ by axial symmetry,}\\
\phi_r(1, z, t) &= 0 \text{ for } 0 \leqslant z \leqslant f(1, t),\\
\phi_z(r, 0, t) &= 0 \text{ for } 0 \leqslant r \leqslant 1.
\end{aligned}$$

On the free surface, the motion is related to the velocity potential by $\mathrm{d}r/\mathrm{d}t = \phi_r$ and $\mathrm{d}z/\mathrm{d}t = \phi_z$. In addition (see reference in paragraph 3 of this chapter) we impose Bernoulli's equation with surface tension taken into account on the free surface, thus

$$\begin{aligned}
\phi_t &= (a(t)/(1 + B^{-1}))\,(z - H) - (\phi_r^2 + \phi_z^2)/2\\
&+ (1 + B)^{-1}\left(\frac{f_r}{r(1 + f_r^2)^{1/2}} + \frac{f_{rr}}{(1 + f_r^2)^{3/2}}\right)
\end{aligned}$$

on the free surface $z = f(r, t)$, for $0 < r < 1$, $t \geqslant 0$, where B^{-1} is a surface tension coefficient and $a(t)$ represents the acceleration of the cylinder along its axis. The constant H is the average height of the fluid, which is assumed to have a constant volume.

We impose the following initial conditions:

$z = f(r, 0) = f_0(r)$, the initial shape of the free surface for $0 \leqslant r \leqslant 1$,

$\phi(r, z, 0) = 0$ for $0 \leqslant r \leqslant 1, 0 \leqslant z \leqslant f_0(r)$, and $a(0) = 0$.

The terms involving f_r and f_{rr} in the nonlinear partial differential equation (Bernoulli's equation) above are the principal radii of curvature of the free surface, which determine the surface tension forces.

The method to be described involves developing a series solution for the velocity potential and determining the time-dependent coefficients by imposing the free surface conditions.

We begin with the separation of variables method for the velocity potential equation. If we put $\phi(r, z, t) = C(t)R(r)Z(z)$, then Laplace's equation in cylindrical coordinates leads to the ordinary differential equations $R'' + (1/r)R' + kR = 0$, and $Z'' - kZ = 0$, with boundary conditions $R'(0) = R'(1) = 0$ and $Z'(0) = 0$. This set of equations and boundary conditions has a countable number of solutions which are linearly independent. The velocity potential can be represented by

$$\phi(r, z, t) = \sum_{n=0}^{\infty} C_n(t) J_0(\lambda_n r) \frac{\cosh(\lambda_n z)}{\cosh(\lambda_n H)},$$

where J_0 is the zero-th order Bessel function and λ_n $(n = 0, 1, \ldots)$ are the roots of the first order Bessel function. Thus, $J_1(\lambda_n) = 0$; that is, $\lambda_0 = 0, \lambda_1 = 3.8317$ $\ldots$, etc.

It remains to determine the time-dependent coefficients $C_n(t)$ and, at the same time, determine the motion of the free surface. It is not an entirely easy matter to do so.

From the expression for $\phi(r, z, t)$, we find that

$$\phi_r(r, z, t) = -\sum_{n=0}^{\infty} \lambda_n C_n(t) J_1(\lambda_n r) \frac{\cosh(\lambda_n z)}{\cosh(\lambda_n H)} \quad \text{and}$$

$$\phi_z(r, z, t) = \sum_{n=0}^{\infty} \lambda_n C_n(t) J_0(\lambda_n r) \frac{\sinh(\lambda_n z)}{\cosh(\lambda_n H)}.$$

We also find that

$$\phi_t(r, z, t) = \sum_{n=0}^{\infty} C_n'(t) J_0(\lambda_n r) \frac{\cosh(\lambda_n z)}{\cosh(\lambda_n H)}.$$

Suppose we now define

$$F_n(r, t) = J_0(\lambda_n r) \frac{\cosh(\lambda_n f(r, t))}{\cosh(\lambda_n H)}, \quad n = 0, 1, 2, \ldots,$$

and $B(r, t)$ to be the right hand side of Bernoulli's equation given above as $\phi_t = (a(t)/\ldots\ldots)$, with $z = f(r, t)$.

For ϕ_r and ϕ_z in the expression for $B(r, t)$, we use the two infinite series given above.

With these definitions, Bernoulli's equation can be written in the form

$$\sum_{n=0}^{\infty} C_n'(t) F_n(r, t) = B(r, t) .$$

We next wish to orthonormalize the set of functions $F_0, F_1, F_2, \ldots$ with respect to a suitable inner product, in order to find a system of differential equations for the C_n's. A first attempt, using

$$(F_i, F_j) = \int_0^1 F_i(r, t) F_j(r, t) \mathrm{d}r$$

led to numerical difficulties later on in the method. It was then noted that a more reasonable inner product would be

$$(F_i, F_j) = \int_0^1 r F_i(r, t) F_j(r, t) \mathrm{d}r ,$$

since, for a fixed increment in radius, the increment in volume of fluid in a cylinder increases linearly with the radius. With this inner product, we used the Gram-Schmidt process to orthonormalize the functions F_0, F_1, $F_2, \ldots$ to obtain $G_n(r, t)$, for $n = 0, 1, 2, \ldots$, with $(G_i, G_j) = 0$ if i and j are different and $(G_i, G_i) = 1$; see Chapter 6.

We can write

$$F_n(r, t) = \sum_{m=0}^{n} b_{nm}(t) G_m(r, t) ,$$

where $b_{nm}(t)$ can be expressed in terms of the time-dependent inner products (F_i, F_j); see Chapter 6.

We now truncate the series for Bernoulli's equation, as written in terms of the functions F_n, at $n = N$, substitute the expressions for F_n in terms of the G_m, and take the inner product of both sides with each of the functions $G_0, G_1, \ldots, G_N$. In this way, we obtain the equations

$$\sum_{n=0}^{N} b_{n0}(t) C_n'(t) = (B, G_0)$$

$$\sum_{n=1}^{N} b_{n1}(t) C_n'(t) = (B, G_1)$$

$$\cdots\cdots\cdots\cdots\cdots\cdots$$

$$b_{NN}(t) C_N'(t) = (B, G_N) .$$

We can solve the system explicitly for $C_N'(t), C_{N-1}'(t), \ldots, C_0'(t)$ by back-substitution proceeding from the last equation backwards to the first. The $(N+1)(N+2)/2$ definite integrals which arise as inner products occurring in this system of equations are functions of $C_n(t), n = 0, 1, \ldots, N$.

We next truncate the series for the velocity potential at $n = N$, along with the series for ϕ_r and ϕ_z. We obtain, in this way, a system of first-order ordinary differential equations to solve for the $C_n(t)$. These equations, together with the equations $\mathrm{d}r/\mathrm{d}t = \phi_r$ and $\mathrm{d}f/\mathrm{d}t = \phi_z$ for a finite number of points $(r_m(t), f_m(t))$ on the free surface, can be written as a first-order autonomous system. The initial conditions are determined by the assumption that the fluid is initially at rest and by a knowledge of the initial shape of the free surface.

We set $C_n(0) = 0, n = 0, 1, 2, \ldots, N$, and choose some distribution of points on the given initial free surface $(r_m(0), f_m(0))$, $m = 1, 2, \ldots, M$. The definite integrals are evaluated by the trapezoidal rule, and the system of $M + N + 1$ differential equations is solved by the modified Euler method to move ahead an increment in time, Δt. In this way, we can determine the time-dependent coefficients $C_n(t)$ and the motion of individual particles on the free surface. The numerical solution at the end of each time step is taken as a new starting condition, and we can continue the solution for another time step.

The method outlined as the advantage of not requiring the surface to be such that $f(r, t)$ is a single valued function of r for a given value of t. As we will see from the results, the method can follow quite complicated motions of the surface.

Numerical experiments with various values of M and N indicated that good results could be obtained in the types of problems of interest with $M = 45$ and $N = 10$, with the initial distribution of points on the free surface chosen so that the points were closer together near the wall of the cylinder.

A difficulty arose in connection with the computation of the radii of curvature in the terms in $B(r, t)$ describing surface tension. At first, the derivatives f_r and f_{rr} were computed using a definite difference scheme. This turned out not to be accurate enough to give meaningful results. As an alternative, the surface tension term, which can be written

$$Tf = \frac{1}{r}\frac{\partial}{\partial r}\left[\frac{r f_r}{(1 + f_r^2)^{1/2}}\right],$$

was considered as defining an operator T, and the first variation of T corresponding to a small variation in f resulting from a small time increment was found. That is, the Fréchet derivative of T was found, resulting in the expression

$$T(f + \delta f) = Tf + \frac{1}{r}\frac{\partial}{\partial r}\left[\frac{r(\delta f)_r}{(1 + f_r^2)^{3/2}}\right] + O((\delta f)_r^2),$$

where $(\delta f)_r = \phi_{zr}\,\Delta t + O(\Delta t^2)$, since $\delta f = \phi_z \Delta t + O(\Delta t^2)$. Using this method for computing the surface tension term, we can obtain sufficiently accurate results by taking Δt small enough. Values of Δt actually used were in the range 0.01 to 0.03 seconds. For hemispherical initial surface shapes, we can rewrite Bernoulli's equation so that Tf is initially zero; see the reference at the start of this chapter.

The complete set of equations needed to apply the method to specific numerical examples was programmed in FORTRAN for the IBM 7094 computer in 1964 by L. M. Perko. Some results obtained using the program will now be described.

Since the results will be shown in graphical form, we begin with a sketch of the geometry of the problem as shown in Fig. 1.

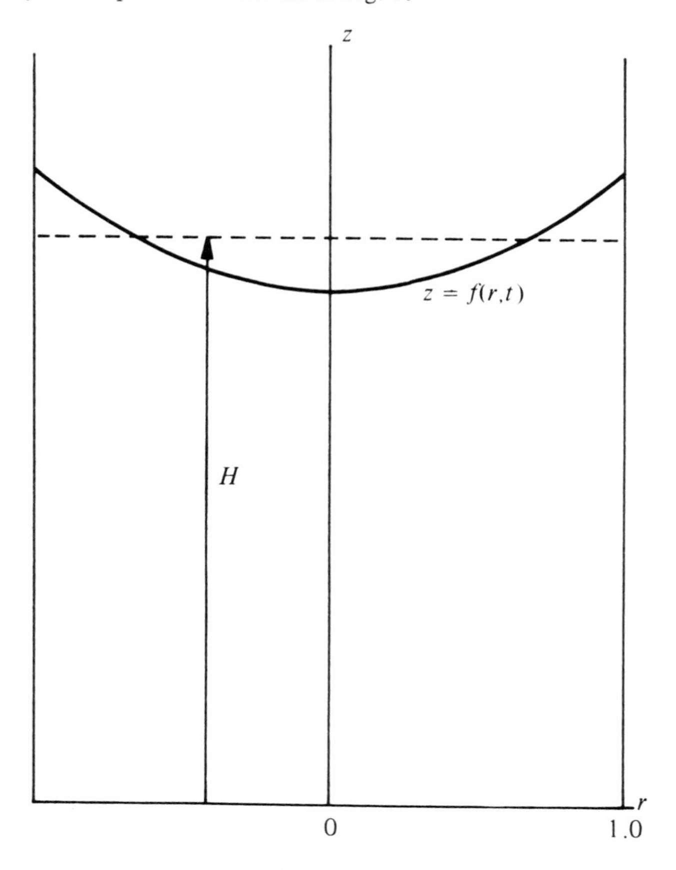

Fig. 1 Coordinate system and hemispherical initial shape

In Fig.s 2–6, with $a(t) = 1.0$ for $t > 0$, the cylinder is being accelerated away from the free surface. This is similar to the case of an upside-down cylindrical container at the earth's surface, gravity acting to pull the liquid out of the container.

In Figs. 7–9, with $a(t) = -1.0$, the cylinder is being accelerated towards the free surface, as with gravity acting to keep the liquid in the container, when right-side up.

Most of the cases reported in the figures begin with a hemispherical shape of radius $R = \sec \theta_0$, for various values of θ_0, which is the angle which the surface makes initially with the wall.

The average initial height H, given in feet, and the actual radius r_0 (in feet) have been normalized by dividing by r_0; time has been divided by $(r_0/g_0)^{1/2}$ and the velocity potential by $(r_0^3 g_0)^{1/2}$. Laplace's equation and the boundary conditions remain invariant under this normalization. Since the fluid is assumed incompressible, the volume should not change. During the computations, we monitored the volume of the fluid as a measure of the accuracy of the computed motion of the free surface. With the choices of M and N used, $M = 45$ and

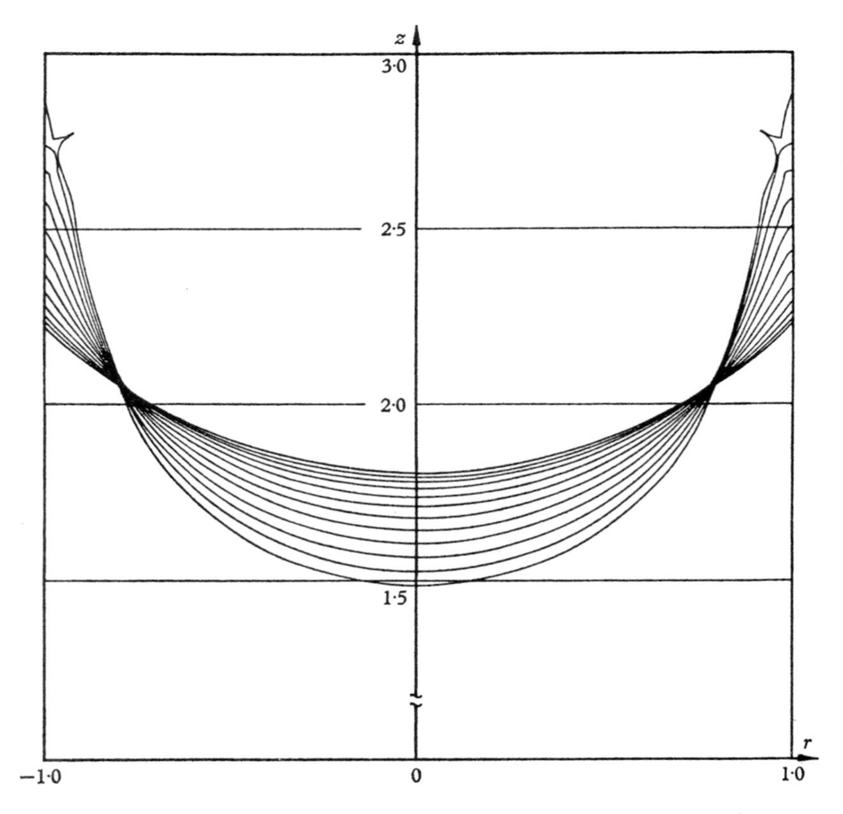

Fig. 2 Hemispherical initial shape with $H = 2.0, a(t) = +1, \beta = 0, \theta_0 = 45°$, and $\Delta t = 0.0177 r_0^{1/2}$ sec

$N = 10$, the volume varied by less than 0.01% per time step for the largest time step used, and still less for smaller time steps.

Varying degrees of surface tension were tried from $\beta = B^{-1} = 0$ (no surface tension) to $\beta = 0.05$ (an example of which would be water in a 0.8 inch diameter cylinder).

The method was even able to handle a case with a nearly flat initial shape with a meniscus at the wall, so that the free surface has initial contact angle zero at the wall; see Fig. 6.

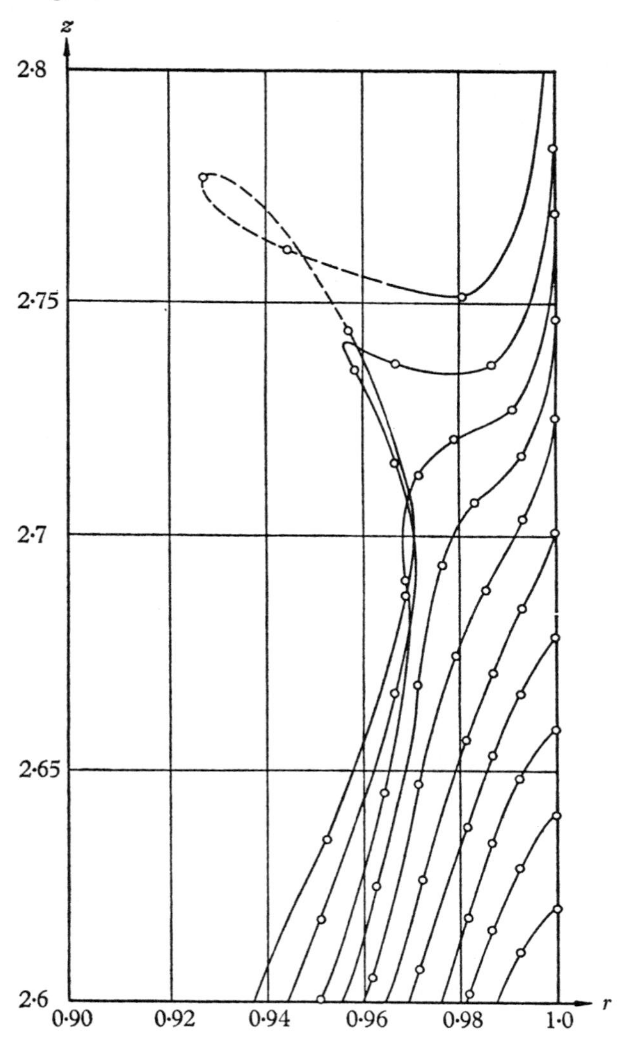

Fig. 3 The breakers at the wall for the case in Fig. 2

Figs. 2 and 5–9 are reproductions of piecewise linear plots made by the SC 4020 plotter directly from computer results. As an indication of computer time required, the total time for the computation of the fluid motion shown in Fig. 2 was 6 minutes on the IBM 7094. Only every fifth computed shape is shown in the figure, for clarity.

Note that, in Fig. 2, breakers developed at the wall of the container. These are shown in more detail in Fig. 3. In further studies, not shown, it was observed

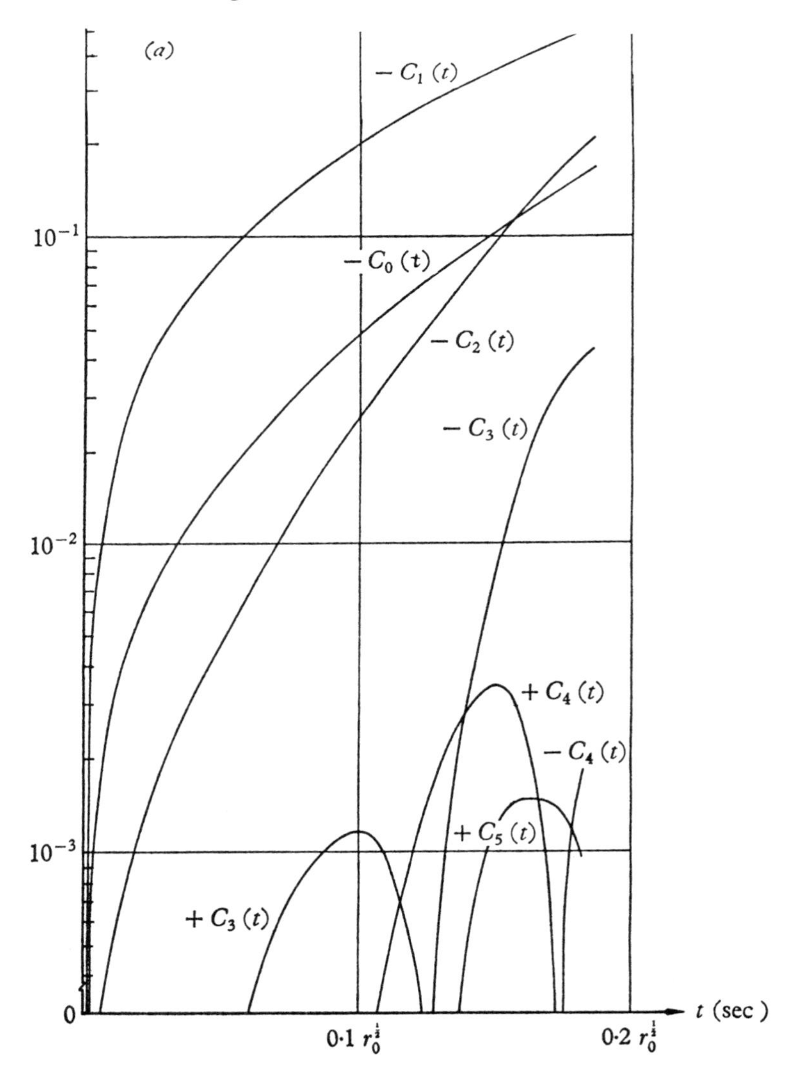

Fig. 4 The time variation of the

that varying the time step for the case shown in Fig. 2 had little or no effect on the overall behavior of the computed motion of the fluid. The breakers appeared at almost exactly the same time t when different time steps were used. The time at which the breakers occurred was, however, found to be very sensitive to small changes in the initial shape. A deviation in the initial hemispherical shape of Fig. 2 by 2% at two or three points on the surface caused a 20% decrease in the time at which breakers appeared. Also the time at which breakers appeared was smaller for smaller contact angles θ_0.

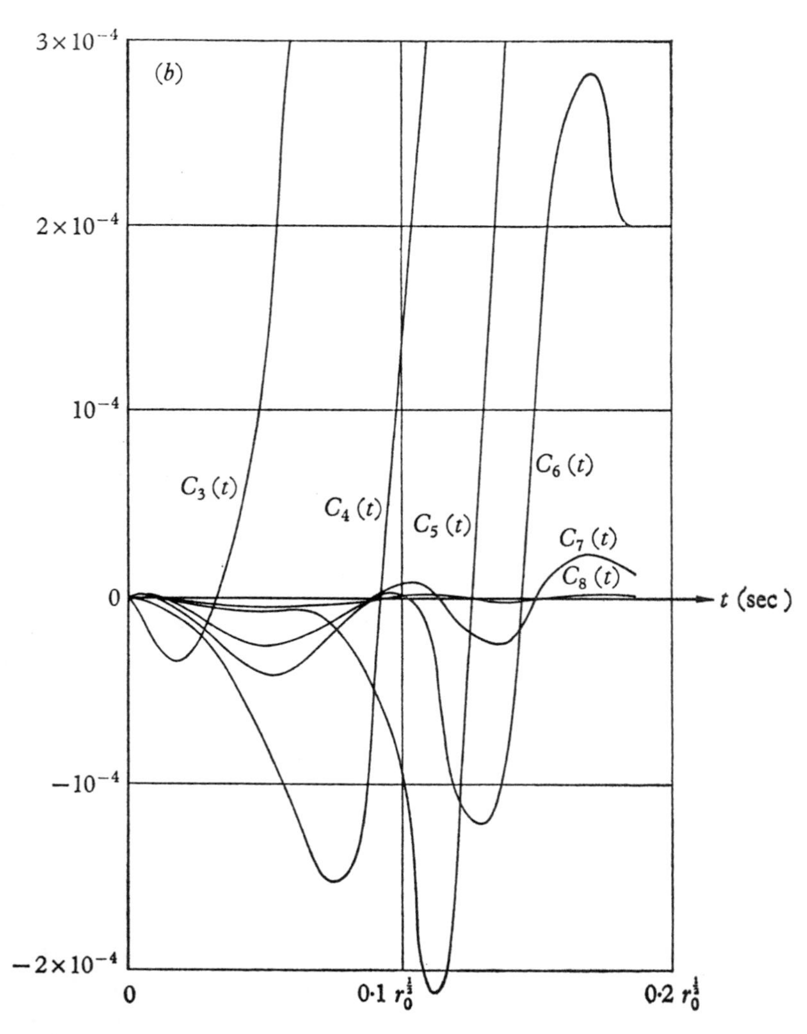

coefficients for the case in Fig. 2

The qualitative behavior was much the same as in Fig. 2 for different volumes of fluid, although a bottom effect was observed in the case of smaller volumes. The time at which breakers appeared did not vary much for different volumes.

The results shown in Fig. 2 are for the case of no surface tension, $\beta = 0$. Surface tension had a smoothing effect which eliminated the breakers for sufficiently large β, see Fig. 5, which differ from the case studied in Fig. 2 only by the introduction of a small surface tension coefficient, $\beta = 0.005$.

To avoid undamped oscillations in the surface tension term arising from overcorrections occurring at places where there was large curvature developing, it was necessary to use variable time steps to maintain a small growth of the surface tension term.

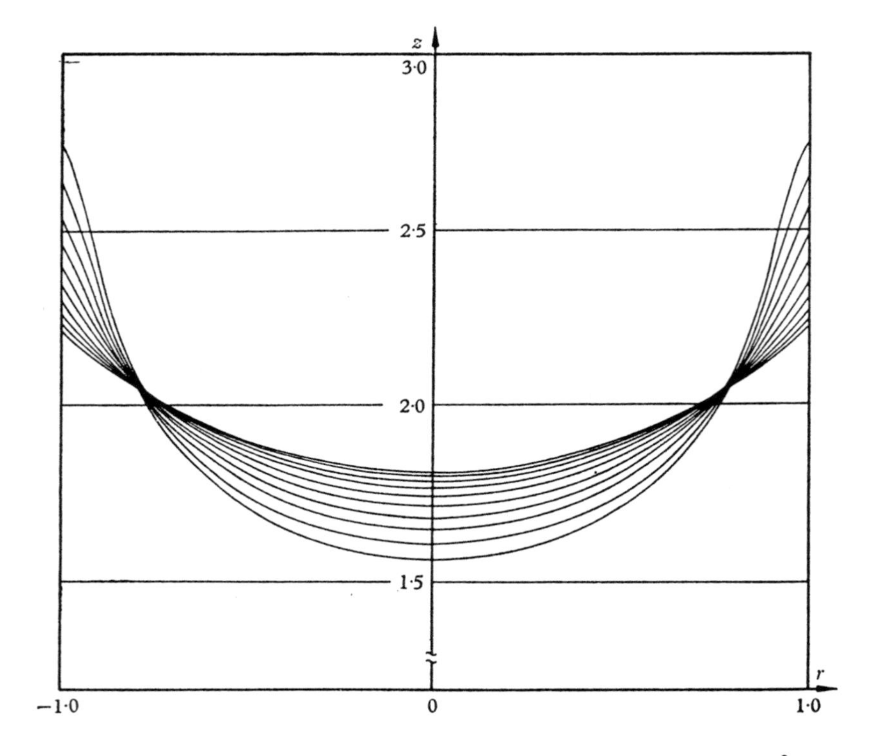

Fig. 5 Hemispherical initial shape with $H = 2.0, a(t) = +1, \beta = 0.005, \theta_0 = 45°$, and $\Delta t = 0.0177 r_0^{1/2}$ sec

In Fig. 4, we can see the time-dependent coefficients $C_n(t)$ plotted for the case studied in Fig. 2. In further studies (not shown) it was observed that increasing the number of terms N beyond 10 did not affect the values of the first ten terms significantly. In particular, $C_8(t)$, the last term plotted, is quite small. The series does seem to be converging.

The time steps indicated in the captions to the figures are the initial time steps. As mentioned, variable time steps were used to maintain small changes in the volume (as an indication of accuracy) and to maintain small changes in the surface tension terms.

In Fig. 6, we consider the case of an initial shape which is flat except for a meniscus near the wall. With surface tension present, the program was able to follow the motion of the surface until the time step became prohibitively small. Without surface tension, the breakers appeared near the wall almost immediately.

A case with acceleration growing linearly with time was considered (but is not shown), with $a(t) = kt$, and with a hemispherical initial shape and no surface tension. The results were qualitatively the same as in Fig. 2; the time at which breakers occurred increased with decreasing k.

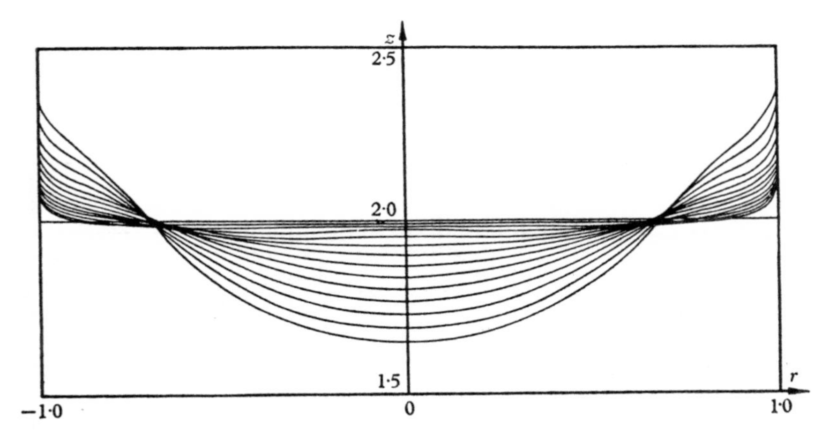

Fig. 6 Flat initial shape with meniscus at the wall with $H = 2.0, a/t = +1, \beta = 0.05$, $\theta_0 = 45°$, and $\Delta t = 0.0350 r_0^{1/2}$ sec

In Figs. 7–9, we consider some cases when $a(t) = -1$, and obtain some interesting kinds of oscillations of the free surface.

The relative magnitudes of surface tension, acceleration, and the initial contact angle determined whether or not a splash developed on the surface.

In Fig. 7, with no surface tension and an initial contact angle of 45°, a crown-shaped splash developed at the center of the fluid.

It was found that with the initial angle equal to 45° and a small surface tension effect, $\beta = 0.005$, similar to water in a 2.5 inch cylinder, the surface goes through one and a half oscillations before a splash starts to develop. With the same initial contact angle, but a larger surface tension, $\beta = 0.05$, no splash occurs, and several complete oscillations of the surface were computed. Half of the first oscillation is shown in Fig. 8. It is interesting to note that the average period of oscillation in this case, $T = 0.485 r_0^{1/2}$, was slightly larger than that

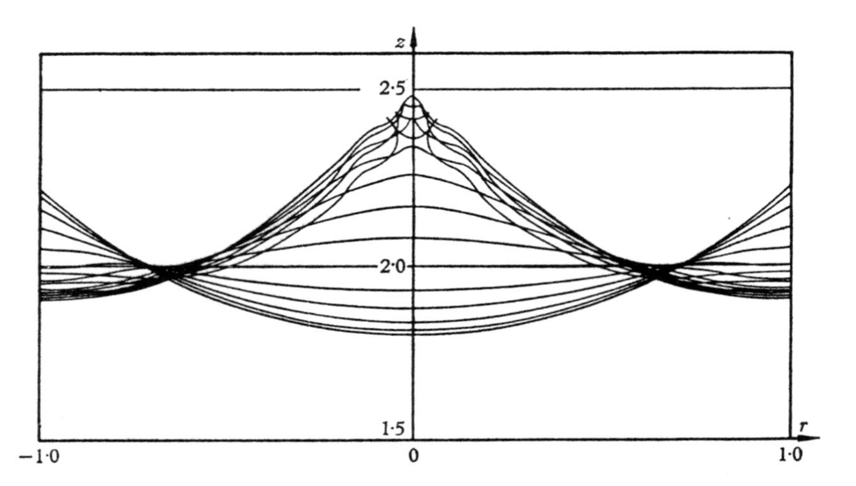

Fig. 7 Hemispherical initial shape with $H = 2.0, a(t) = -1, \beta = 0, \theta_0 = 45°$, and $\Delta t = 0.0266 r_0^{1/2}$ sec

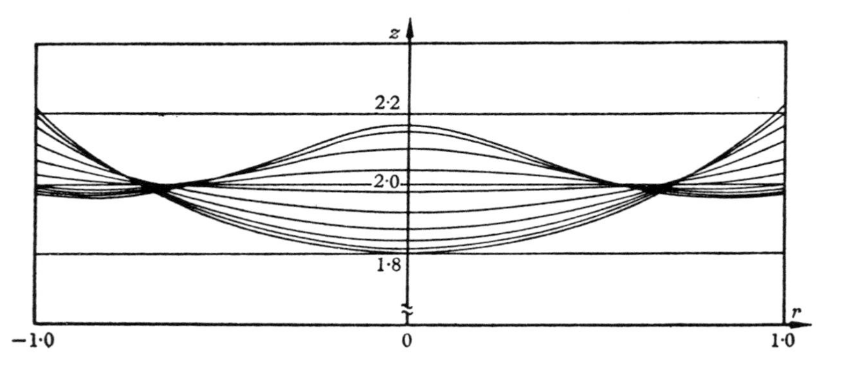

Fig. 8 Hemispherical initial shape with $H = 2.0, a(t) = -1, \beta = 0.05, \theta_0 = 45°$, and $\Delta t = 0.0266 r_0^{1/2}$ sec

given by the linear theory for small oscillations, namely $T = 0.434 r_0^{1/2}$. Some crude experiments were performed by the author, timing the period of oscillations of water in cylinders of small diameter. The results were in rough agreement with computed periods in cases such as that of Fig. 8, which corresponds to water in a cylinder of diameter 0.8 inches.

In the case shown in Fig. 9, with a smaller initial angle, a splash develops owing to the larger amount of potential energy in the initial surface shape, even with the same surface tension as in Fig. 8. The motion is shown in two parts. In Fig. 9(a), the center of the fluid rises up as the fluid goes down at the wall. In Fig. 9(b), the fluid at the wall goes back down, while a crown-shaped splash develops in the center.

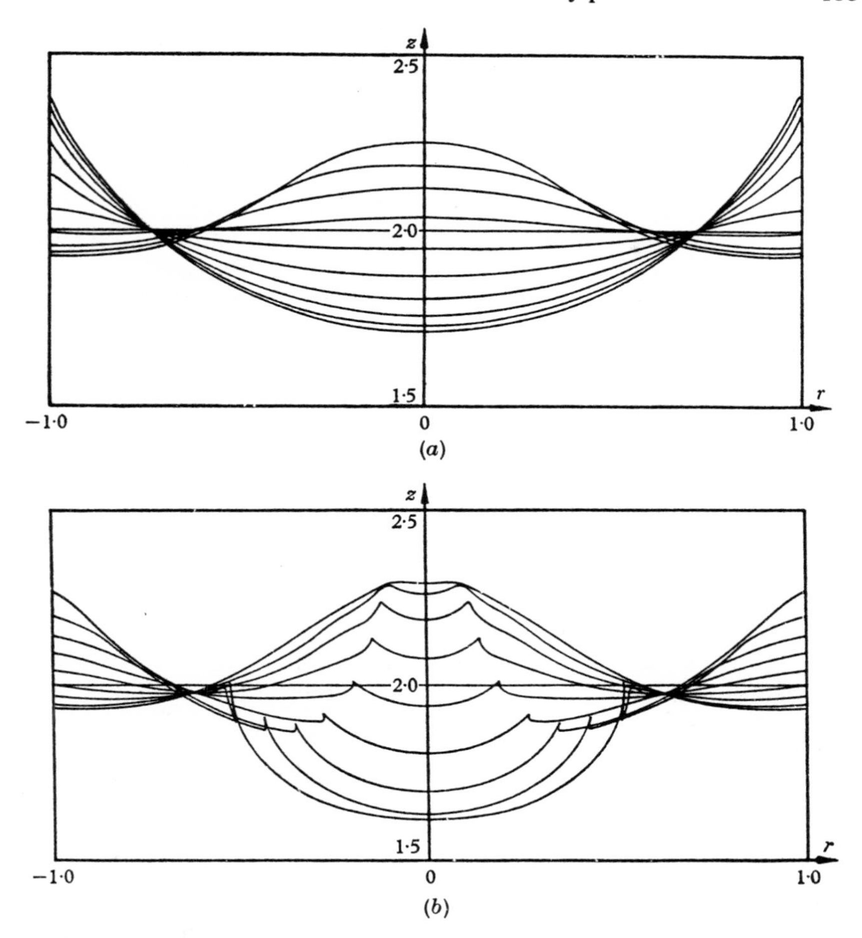

Fig. 9 (a) Hemispherical initial shape with $H = 2.0$, $a(t) = -1$, $\beta = 0.05$, $\theta_0 = 15°$, and $\Delta t = 0.0266 r_0^{1/2}$ sec
(b) A continuation of Fig. 9(a)

Index